MATHS
NOW!
· · · · · · · · · · · ·
GET THE POINT!

TONY &
MARY ELLEN BELL

JOHN MURRAY

Photo acknowledgements

Cover reproduced by courtesy of ZEFA; **p.5** John Townson/Creation; **p.31** Last Resort Picture Library; **p.44** Sporting Pictures (UK) Ltd; **p.53** Tony & Mary Ellen Bell; **p.54** *all* Tony and Mary Ellen Bell; **p.61** *both* Last Resort Picture Library; **p.62** *both* Last Resort Picture Library; **p.82** *both* John Townson/Creation; **p.83** ZEFA; **p.102** Last Resort Picture Library; **p.118** Last Resort Picture Library; **p.165** John Townson/Creation; **p.205** Mary Evans Picture Library; **p.206** BBC Hulton Picture Library; **p.208–10** *all* Tony and Mary Ellen Bell.

The authors would like to thank Kim O'Driscoll, researcher in low attainment in mathematics, University of Strathclyde, and all the schools and teachers throughout the country who helped in the development of this book.

© Tony and Mary Ellen Bell 1998

First published in 1998
by John Murray (Publishers) Ltd
50 Albemarle Street
London W1X 4BD

Reprinted 2000

Layouts by Stephen Rowling/unQualified Design.
Artwork by Tom Cross, Mike Humphries and Janek Matysiak.
Cover design by John Townson/Creation.

Typeset in 12/14pt Times by Wearset, Boldon, Tyne and Wear.
Printed and bound by G. Canale, Italy.

A CIP catalogue record for this book is available from the British Library.

ISBN 0 7195 7278 9

Teacher's Resource File 2 ISBN 0 7195 7279 7

Contents

How to use this book

This maths book is planned to help you understand and enjoy maths. You will be able to gain points which you will collect on a sheet so that you can see how well you are doing. You can swap these points for rewards.

In this book you will meet some symbols. They will tell you what you need and what to do. Here they are.

Work with a partner

Work in a group

See your teacher

Fetch equipment

Take a test

Stop and think

Copy and complete

 When you **copy and complete**, replace a box ☐ with a number

and a line _____ with a word or words.

Sometimes you are given an example to show you how to start. These are always written in red, like this.

We hope that you will enjoy this book.

Number

Addition facts to 20

Unit 1 words

more	less than	total
odd	even	add
plus	eleven	twelve
between	cost	nearer to

Remember

Examples are shown in red.

means copy and complete.

 You need

- a set of Unit 1 vocabulary Snap cards.

 Play a game of Snap to help you learn the words.

 Try the **word test** to get some points.

1 Match the ticket to the coin.

Sort the coins

2 a) Make 2 lists to sort the coins.
List the **copper** coins.
List the **silver** coins.

Copper	Silver
①p	⑤p

b) Make 2 lists to sort the coins.
List the coins that are **circles**.
List the coins that are **not circles**.

Circles	Not circles

3   Try Worksheet 1 *All change*.

Sort the shopping

4 a) Make 2 lists to sort the shopping.
List the things that cost **more than** £1.
List the things that cost **less than** £1.

Cost more than £1	Cost less than £1
CD	

5 (You can use 1p, 2p and 5p coins to help you.)
How many different ways can each of you make 10p?

Did you find more ways than your partner?

6 Draw the pairs of bags that have the **same amount** of money in them.
Put an = sign between the bags.

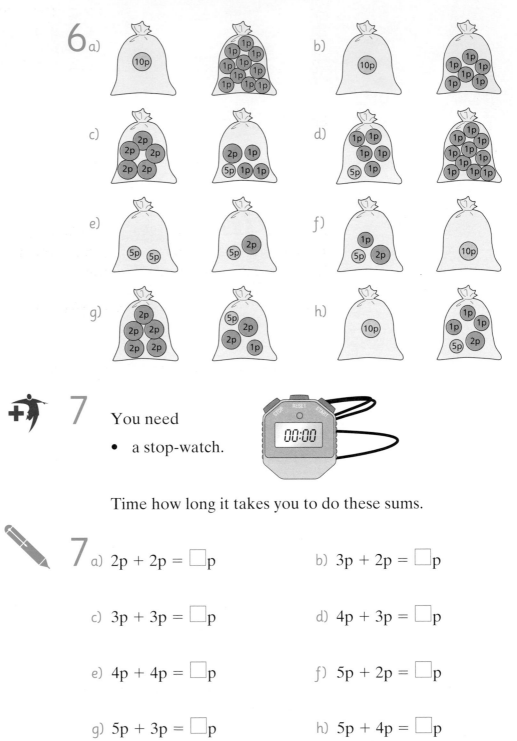

7 You need

• a stop-watch.

Time how long it takes you to do these sums.

7 a) 2p + 2p = ☐p   b) 3p + 2p = ☐p

c) 3p + 3p = ☐p   d) 4p + 3p = ☐p

e) 4p + 4p = ☐p   f) 5p + 2p = ☐p

g) 5p + 3p = ☐p   h) 5p + 4p = ☐p

i) $5p + 5p = \boxed{}p$　　　　j) $6p + 2p = \boxed{}p$

k) $6p + 3p = \boxed{}p$　　　　l) $6p + 4p = \boxed{}p$

m) $7p + 2p = \boxed{}p$　　　　n) $7p + 3p = \boxed{}p$

o) $8p + 2p = \boxed{}p$

It took me $\boxed{}$ minutes to do these sums.

8 (You can use 1p, 2p and 5p coins to help you.)
How many different ways can each of you make 20p?
You must use a 10p coin each time.

Did you find more ways than your partner?

9 Try Worksheet 2 *Which hand?*

10 Copy the number lines and the arrows, or use
Worksheet 3. Then complete the sums.

10 a) 0 1 2 3 4 5 6 7 8 9 10 11 12 13 14 15 16 17 18 19 20

$10 + 1 = \boxed{}$

b) 0 1 2 3 4 5 6 7 8 9 10 11 12 13 14 15 16 17 18 19 20

ten **and** three $= \boxed{}$

c) 0 1 2 3 4 5 6 7 8 9 10 11 12 13 14 15 16 17 18 19 20

9 **plus** three $= \boxed{}$

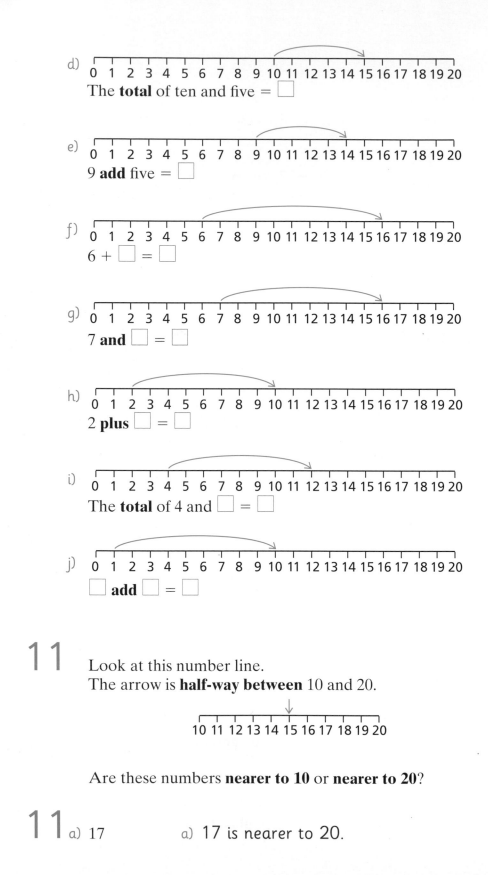

d) The **total** of ten and five = ☐

e) 9 **add** five = ☐

f) 6 + ☐ = ☐

g) 7 **and** ☐ = ☐

h) 2 **plus** ☐ = ☐

i) The **total** of 4 and ☐ = ☐

j) ☐ **add** ☐ = ☐

11 Look at this number line.
The arrow is **half-way between** 10 and 20.

Are these numbers **nearer to 10** or **nearer to 20**?

11 a) 17   a) **17** is nearer to 20.

b) 19   c) 13   d) 16   e) 14

f) 11   g) 18   h) 12

12 Meena and her friends buy bus tickets.

Meena buys a ticket for herself  and a ticket for **1** friend.
How much does she spend if she buys a ticket for herself **and**:

12
a) Kev?   b) Joe?   c) Jung?   d) Emma?

a) Meena + Kev
10p + 8p = 18p

e) Ali?   f) Jade?   g) Tom?   h) Kelly?

13 Look again at the picture in **Question 12**.
Kelly buys a ticket for herself ⟨9p⟩ and a ticket for **1** friend.
How much does she spend if she buys a ticket for herself **and**:

13 a) Kev?   b) Joe?   c) Jung?   d) Emma?

a) Kelly + Kev
   9p + 8p = 17p

e) Ali?   f) Jade?   g) Tom?   h) Meena?

14 Look at your answers to **Questions 12** and **13**.

12a)   answer 18p

13a)   answer 17p

What do you notice about them all?

15 Can you think of a quick way to add **9** to numbers up to 10?
Tell your teacher when you think you have found a way.

16 Try Worksheet 4 *Stack the shelves*.

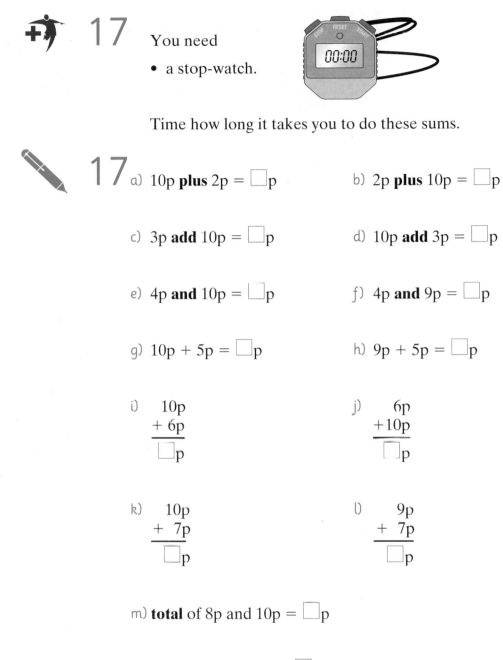

17 You need
• a stop-watch.

Time how long it takes you to do these sums.

17

a) 10p **plus** 2p = ☐p   b) 2p **plus** 10p = ☐p

c) 3p **add** 10p = ☐p   d) 10p **add** 3p = ☐p

e) 4p **and** 10p = ☐p   f) 4p **and** 9p = ☐p

g) 10p + 5p = ☐p   h) 9p + 5p = ☐p

i)   10p
  + 6p
   ☐p

j)   6p
  +10p
   ☐p

k)   10p
  + 7p
   ☐p

l)   9p
  + 7p
   ☐p

m) **total** of 8p and 10p = ☐p

n) **total** of 9p and 8p = ☐p

It took me ☐ minutes to do these sums.

18 How many dots on each domino?

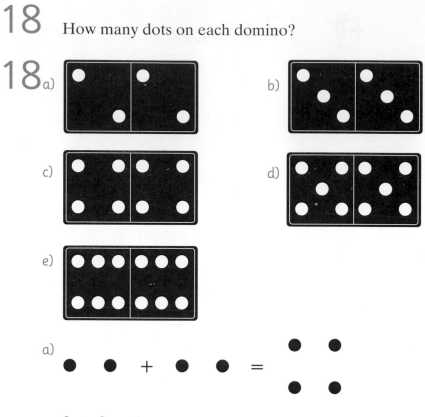

18 a) b)

c) d)

e)

a)

$$2 + 2 = 4$$

 Odds and evens

> **Remember**
>
> An **even number** can be made into **pairs**.
>
>
>
> An **odd number** has an **odd one** left over.

19 a) Write down all the **odd numbers** between 0 and 10.
Double each of them.
Are your answers **odd** or **even**?

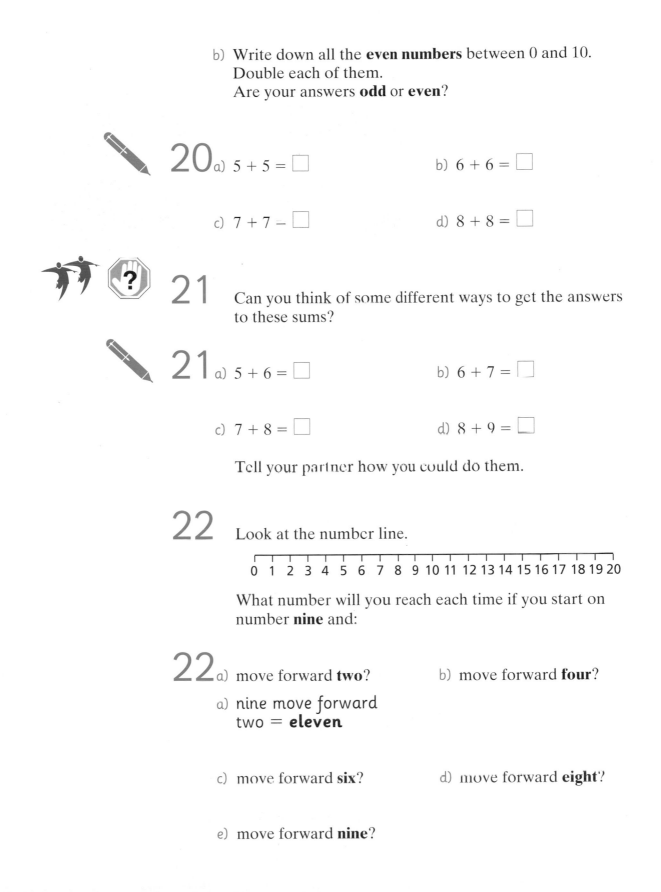

b) Write down all the **even numbers** between 0 and 10.
Double each of them.
Are your answers **odd** or **even**?

20 a) $5 + 5 = \square$ b) $6 + 6 = \square$

c) $7 + 7 = \square$ d) $8 + 8 = \square$

21 Can you think of some different ways to get the answers to these sums?

21 a) $5 + 6 = \square$ b) $6 + 7 = \square$

c) $7 + 8 = \square$ d) $8 + 9 = \square$

Tell your partner how you could do them.

22 Look at the number line.

0 1 2 3 4 5 6 7 8 9 10 11 12 13 14 15 16 17 18 19 20

What number will you reach each time if you start on number **nine** and:

22 a) move forward **two**?   b) move forward **four**?

a) nine move forward two = **eleven**

c) move forward **six**?   d) move forward **eight**?

e) move forward **nine**?

23 What number will you reach each time if you start on number **seven** and:

23 a) move on **four**? b) move on **five**?

c) move on **eight**? d) move on **nine**?

e) move on **seven**?

24 What number will you reach if you:

24 a) start on number **five** and add **five**?

b) start on number **six** and add **six**?

c) start on number **seven** and add **seven**?

d) start on number **eight** and add **eight**?

e) start on number **nine** and add **nine**?

25 a) $4 + \square = 11$ b) $\square + 4 = 11$

c) $7 + \square = 10$ d) $\square + 7 = 10$

e) $\square + 7 = 13$ f) $7 + \square = 13$

g) $5 + \square = 12$ h) $\square + 5 = 12$

26 Ron collects hats.

He has **8** hats already. How many would he have with:

26 a) 2 more hats?   b) 7 more hats?

c) 3 more hats?   d) 6 more hats?

e) 5 more hats?   f) 8 more hats?

g) 4 more hats?   h) 9 more hats?

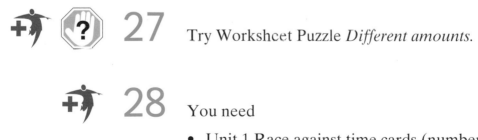

27 Try Worksheet Puzzle *Different amounts.*

28 You need

- Unit 1 Race against time cards (numbers and words)
- your 'My maths record' sheet.

Race against time

1 Sort the race cards – this side up. | 7 + 7 |

2 Take the cards 1 at a time.

3 Answer as quickly as you can.

4 Did you get it right?
Look at the other side of the card. | fourteen |

5 When you get all the answers correct,
ask a friend to test you.

6 Now **Race against time**

Go for points!

Ask your teacher to test and time you.

Remember
3 errors – 1 point
2 errors – 2 points
1 error – 3 points
0 errors – 5 points
Answer in 1 minute with 0 errors – 7 points

Now try Unit 1 Test.

Review 1

1 Write in **numbers**.

1 a) seventy   b) thirteen

c) ninety-six   d) seventeen

2 Write in **words**.

2 a) 80   b) 18

c) 50   d) 9

3 Write in order, **smallest** number first.

9   27   80   2   31   13

4 How many tallies?

4 a) 十十 十十 lll   b) 十十 ll

c) 十十 十十 十十 十十 l

5 Copy each group of tallies. Add **3 more** to each group. Write how many tallies there are **altogether**.

> **Remember**
>
> Write tallies in groups of 5.

5 a) ‖‖‖ ‖‖‖ ‖

   a) ‖‖‖ ‖‖‖ ‖‖‖ 15 tallies altogether

   b) ‖‖‖ ‖‖‖ ‖‖‖ ‖‖‖

   c) ‖‖   d) ‖‖‖ ‖‖

6 How many people? 👤 = 1 person

6 a) 👤👤   = ☐ people

   b) 👤👤👤👤👤   = ☐ people

   c) 👤👤👤👤👤👤👤👤👤 = ☐ people

2 Handling data

Block graphs and pictograms

Unit 2 words

most	altogether	thirteen
fifteen	twenty	thirty
forty	pictogram	block graph
tally	cost	fewer

Remember

Examples are shown in red.

Means copy and complete.

You need

• a set of Unit 2 vocabulary Snap cards.

Play a game of Snap to help you learn the words.

Try the **word test** to get some points.

1 Pupils in tutor group 9B found out what time the people in their tutor group go to bed.

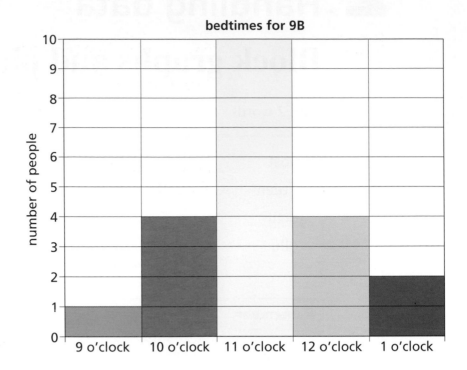

Most people went to bed at 11 o'clock.

1 a) ☐ people went to bed at 11 o'clock.

b) The numbers at the **side** of the graph go up to number ☐.

c) Look at your answers to **Questions 1a)** and **b)**.
What do you notice?

d) People chose ☐ different bedtimes.

e) There are ☐ columns.

f) Look at your answers to **Questions 1d)** and **e)**.
What do you notice?

g) Would the graph look different if a **different** group of people were asked?
You could think about:

- age
- work
- time of year
- weekday or weekend
- telling the truth . . .

2   Try Worksheet 1 *Cut and paste.*

3   This tally table shows which sorts of soap powder a group of people use.

Remember
1 tally mark stands for **1** person.

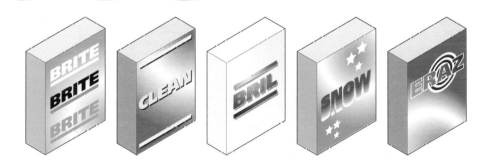

Soap powder	Tally marks
Eraz	ⵊⵊⵊ |||
Clean	ⵊⵊⵊ	
Brite	ⵊⵊⵊ ⵊⵊⵊ
Bril	|||
Snow	ⵊⵊⵊ ||

3 a) The biggest number of **tallies** is ☐.

b) The numbers at the **side** of a graph to show soap powder use will go up to ☐.

c) There are ☐ **choices** of soap powder.

d) There will be ☐ **columns** on the graph.

4 Emma and Warren tried to draw a block graph from the tally table in **Question 3**. They tried 4 times.
They made a different mistake each time. Find their mistakes.
Look at these rules to help you.

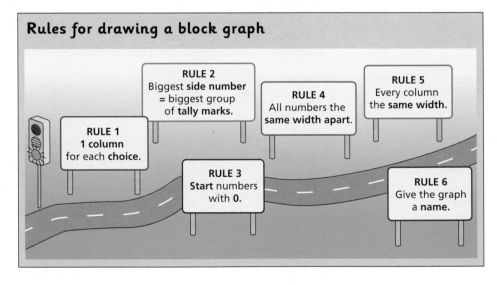

Rules for drawing a block graph

RULE 1
1 column
for each **choice**.

RULE 2
Biggest **side number**
= biggest group
of **tally marks**.

RULE 3
Start numbers
with **0**.

RULE 4
All numbers the
same width apart.

RULE 5
Every column
the **same width**.

RULE 6
Give the graph
a **name**.

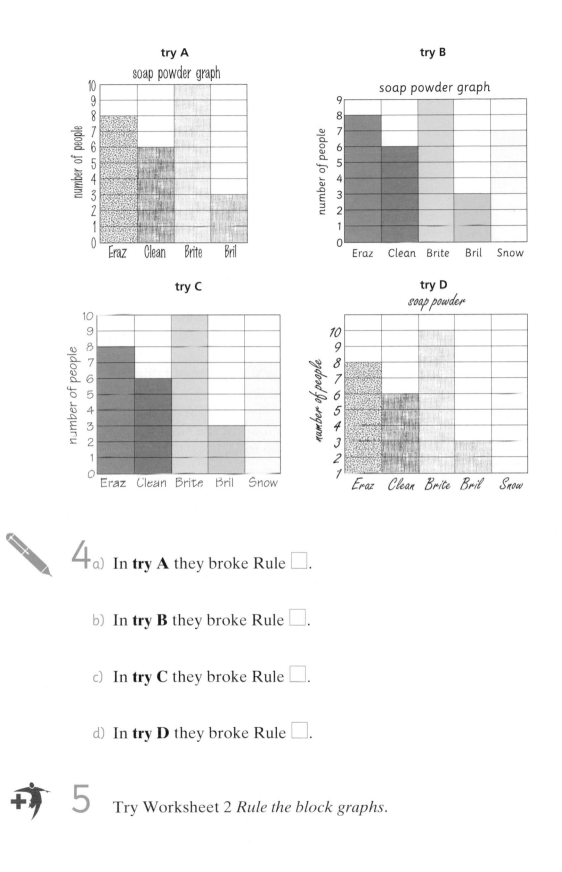

4 a) In **try A** they broke Rule ☐.

b) In **try B** they broke Rule ☐.

c) In **try C** they broke Rule ☐.

d) In **try D** they broke Rule ☐.

5   Try Worksheet 2 *Rule the block graphs*.

6 Here is another tally table. It shows what types of TV programme a group of people watch.

Type of TV programme	Tally marks
comedy	⊩⊩⊩ ⊩⊩⊩
sport	⊩⊩⊩ ⊩⊩⊩
news	⊩⊩⊩ ⊩⊩⊩ ‖
game show	‖‖
film	⊩⊩⊩ ‖‖
soap	‖‖‖

6 a) The biggest number of **tallies** is ☐.

b) The numbers at the **side** of a graph to show TV programmes will go up to ☐.

c) There are ☐ **choices** of programme type.

d) There will be ☐ **columns** on the graph.

e) Use the tally table to draw a block graph.
Use Worksheet 3 (squared paper) or Worksheet 4 to help you.

f) Would the graph look different if a **different** group of people were asked?
You could think about:

- age
- work
- men or women . . .

7 Do your own TV survey in your tutor group.
Use Worksheet 5 to help you. Or make a tally table. Then draw your block graph using Worksheet 3 (squared paper).

Remember the rules.

8 Pupils in tutor group 8A found out about the **crisps** people in their tutor group eat.

8A's favourite crisps

salt and vinegar	🗳🗳🗳🗳🗳🗳🗳🗳
cheese and onion	🗳🗳🗳🗳🗳🗳🗳🗳🗳
plain unsalted	🗳🗳
beef and tomato	🗳🗳🗳🗳🗳🗳🗳🗳
prawn cocktail	🗳🗳🗳🗳

key: 🗳 = 1 packet of crisps

8 a) Which crisps were **most** popular?

b) _____ and _____ were **equally** popular.

c) 🗳 stands for ☐ packet of crisps.

9 These pupils were asked about their favourite colour. This table shows the colours they chose.

Colour	People who chose that colour
red	Len Lee Paul Tracey
blue	Ali Ron Bill Val Ann Jade
orange	Bob Liz Pat Hu Karl Con
green	Sue Sanjay Lan Rashmi
purple	Tom

9 a) Make a tally table.

b) Think of a name for this survey.

10 Jane and Ali tried to draw a pictogram from the tally table in **Question 9**. They tried twice.
They made a different mistake each time. Find their mistakes.
Look at these rules to help you.

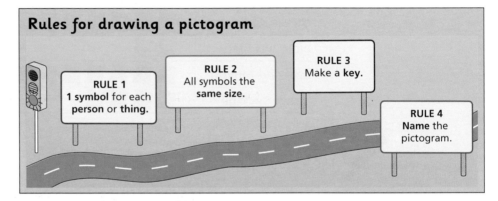

Rules for drawing a pictogram

RULE 1
1 symbol for each **person** or **thing**.

RULE 2
All symbols the **same size**.

RULE 3
Make a **key**.

RULE 4
Name the pictogram.

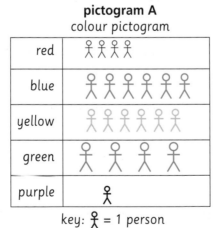

pictogram A
colour pictogram

red	웃 웃 웃 웃
blue	웃 웃 웃 웃 웃 웃
yellow	웃 웃 웃 웃 웃 웃
green	웃 웃 웃 웃
purple	웃

key: 웃 = 1 person

pictogram B
colour pictogram

red	웃 웃 웃 웃
blue	웃 웃 웃 웃 웃 웃
yellow	웃 웃 웃 웃 웃 웃
green	웃 웃 웃 웃
purple	웃

10 a) In **pictogram A** they broke Rule ☐.

b) In **pictogram B** they broke Rule ☐.

11 Pupils in Year 9 found out about the crisps people in their year group eat.

Year 9's favourite crisps

salt and vinegar	🔲🔲🔲🔲🔲🔲🔲🔲🔲
cheese and onion	🔲🔲🔲🔲🔲🔲🔲🔲🔲
plain unsalted	🔲🔲🔲🔲🔲
beef and tomato	🔲🔲🔲🔲🔲🔲🔲🔲
prawn cocktail	🔲🔲🔲🔲

key: 🔲 = 2 packets of crisps

11 a) Did they ask **more** or **fewer** people than the **crisp** survey in Question 8?

They asked _____ people.

b) 🔲 stands for ☐ packets of crisps.

c) ☐ people eat cheese and onion crisps. 🔲

d) ☐ people eat prawn cocktail crisps. 🔲

e) ☐ people eat plain unsalted crisps. 🔲

f) ☐ people eat salt and vinegar crisps. 🔲

g) _____ were the **most** popular crisps.

h) _____ and _____ were **equally** popular.

12 This **extra** flavour of crisps was added to the pictogram:

smoky bacon	

key: = 2 packets of crisps

12 a) What do you think the symbol stands for?

It stands for _____ packet of crisps.

b) ☐ people liked smoky bacon crisps.

13 Try Worksheet 6 *Find the key*.

14 Pupils in Year 7 found out what drink everyone has at break.

Year 7's drinks

milk	
lemonade	
water	
orange	
cola	

key: = 2 drinks

14 a) stands for ☐ drinks.

b) What to you think stands for?

It stands for _____ glass of drink.

c) ☐ glasses of lemonade were drunk.

d) ☐ glasses of milk were drunk.

e) ☐ glasses of orange were drunk.

f) The most popular drink was _____.

15 a) Make up your own keys for things you might want to find out.
Make up a symbol to stand for **1** person or thing.

◼ = 1 CD

b) Make up a symbol to stand for **2** people or things.

◼ = 2 CDs

How would you show **3**, **5**, **7** and **9** things?

16 Do your own **ice-cream** survey in your tutor group. Make a tally table. Then draw your pictogram.
1 symbol = 2 ice-creams.

Now try Unit 2 Test.

Review 2

1 Write in **numbers**.

a) thirty-five   b) twenty   c) thirteen

d) forty   e) ninety   f) forty-four

g) seven   h) eighty-six

2 Write in **words**.

a) 18   b) 81   c) 123   d) 213   e) 301

3 Write in order, **smallest** number first.

41   6   72   14   20   55

4 Copy the numbers. Circle the digit that has the **bigger** value.

a) 17   b) 92   c) 60   d) 48   e) 83

5 What fraction is coloured in each bar, a **half** or a **quarter**?

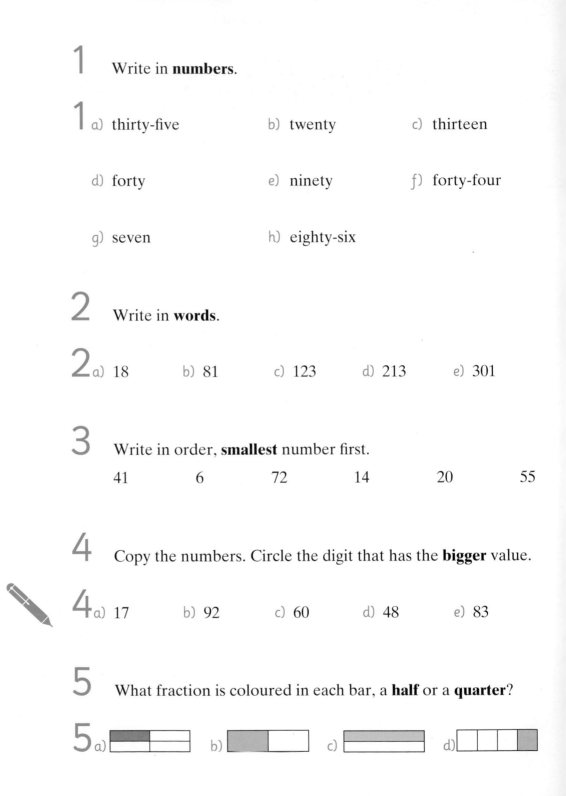

6 a) $\frac{1}{2}$ of 4 sweets = ☐ sweets.   b) $\frac{1}{2}$ of 10 sweets = ☐ sweets.

c) $\frac{1}{2}$ of 8 sweets = ☐ sweets.   d) $\frac{1}{2}$ of 6 sweets = ☐ sweets.

7

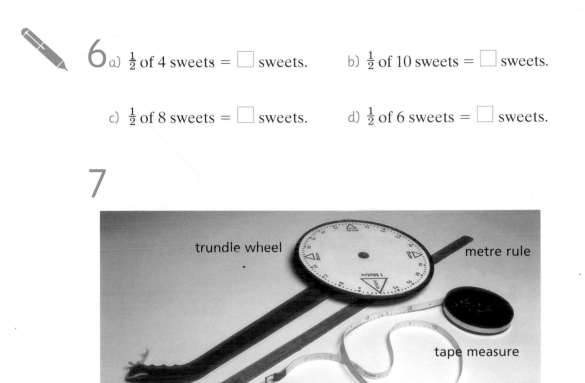

Choose the best measuring tool to measure these things:

7 a) a path

b) a pencil

c) a window

d) your waist.

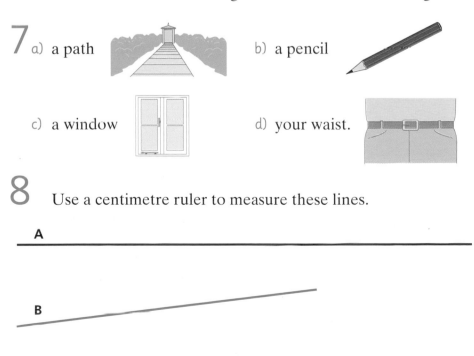

8 Use a centimetre ruler to measure these lines.

A _____

B _____

3: Number

Place value to 200

Unit 3 words

tens	units	roughly
bigger	smaller	worth
value	middle number	hundred
single	between	altogether

Remember

Examples are shown in red.

Means copy and complete.

You need

• a set of Unit 3 vocabulary Snap cards.

Play a game of Snap to help you learn the words.

Try the **word test** to get some points.

1   Change the (1p) coins to (10p) coins.

Remember

Ten 1p coins = one 10p coin.

1 a) Twelve 1p coins = <u>one</u> 10p coin and <u>two</u> 1p coins.

b) Nineteen 1p coins = _____ 10p coin and _____ 1p coins.

c) Twenty 1p coins = _____ 10p coins and _____ 1p coins.

d) Twenty-nine 1p coins = _____ 10p coins and _____ 1p coins.

e) Thirty 1p coins = _____ 10p coins and _____ 1p coins.

f) Thirty-nine 1p coins = _____ 10p coins and _____ 1p coins.

g) Forty 1p coins = _____ 10p coins and _____ 1p coins.

h) Forty-nine 1p coins = _____ 10p coins and _____ 1p coins.

i) Fifty 1p coins = _____ 10p coins and _____ 1p coins.

j) Sixty-eight 1p coins = _____ 10p coins and _____ 1p coins.

k) Eighty-six 1p coins = _____ 10p coins and _____ 1p coins.

l) Ninety-nine 1p coins = _____ 10p coins and _____ 1p coins.

2 Which is worth more, one (1p) coin or one (10p) coin?

a) One _____ coin is **worth more**.

b) One _____ coin has the **bigger value**.

Which is worth less, one (1p) coin or one (10p) coin?

c) One _____ coin is **worth less**.

d) One _____ coin has the **smaller value**.

3 Copy these numbers.
Circle the digit that has the **bigger** value.

a) ①3　　b) 15　　c) 51　　d) 86　　e) 68

f) 47　　g) 74　　h) 39　　i) 93　　j) 90

Say it, write it

4 Play a game of 'Say it, write it'.

Rules for 'Say it, write it'

1 Get a friend to read the numbers in Question 3 aloud.
2 You write the numbers down.
3 Check to see how many you wrote correctly.
4 Change over. Now you read the numbers in Question 3 aloud.
5 Your friend writes the numbers down.
6 Check to see how many numbers your friend wrote correctly.
7 The winner is the player who wrote the most correct numbers.

5 Sam wants to know how many sweets Sue has.
He does not want to know the **exact** number.
He wants to know **roughly** – to the **nearest ten**.

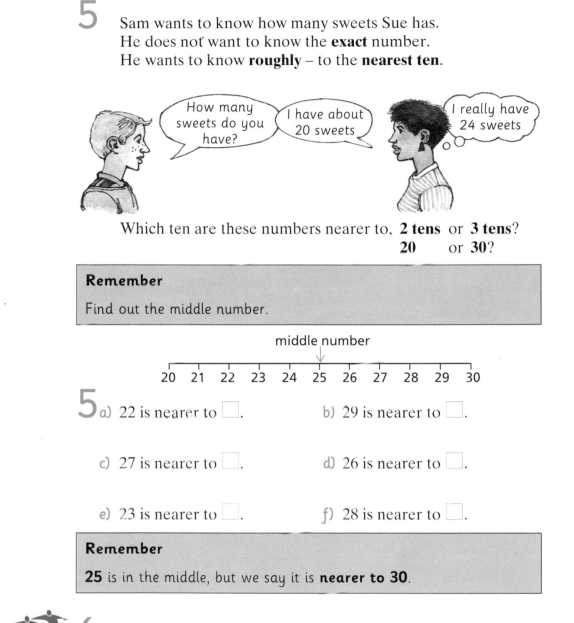

Which ten are these numbers nearer to, **2 tens** or **3 tens**?
20 or **30**?

Remember

Find out the middle number.

middle number
↓
20 21 22 23 24 25 26 27 28 29 30

5 a) 22 is nearer to ☐.   b) 29 is nearer to ☐.

c) 27 is nearer to ☐.   d) 26 is nearer to ☐.

e) 23 is nearer to ☐.   f) 28 is nearer to ☐.

Remember

25 is in the middle, but we say it is **nearer to 30**.

6 Look at this picture of Sam and Sue.

Guess how many different numbers of sweets Sue could
really have.

7 What are these numbers **roughly** – to the **nearest ten**?

7 a) 36 is between ⟦30⟧ and ⟦40⟧.

The middle number is ⟦35⟧.

36 is **roughly** ⟦40⟧.

b) 42 is between ☐ and ☐.

The middle number is ☐.

42 is **roughly** ☐.

c) 58 is between ☐ and ☐.

The middle number is ☐.

58 is **about** ☐.

d) 63 is between ☐ and ☐.

The middle number is ☐.

63 is **about** ☐.

e) 79 is between ☐ and ☐.

The middle number is ☐.

79 is **roughly** ☐.

f) 84 is between ☐ and ☐.

The middle number is ☐.

84 is **about** ☐.

g) 35 is the middle number between ☐ and ☐.

We say 35 is **about** ☐.

h) 65 is the middle number between ☐ and ☐.

We say 65 is **roughly** ☐.

i) 85 is the middle number between ☐ and ☐.

We say 85 is **about** ☐.

j) Can you think of some questions you can ask yourself to help you find a number **roughly** – to the **nearest ten**? Tell your teacher when you have thought of them.

8 Ben's dad has a shop.

He sells biscuits in packets of ten.

He puts some single biscuits
on a plate for people to taste.

How many biscuits are there altogether in each drawing below?

Remember

10 **units** = 1 **ten**

and

10 **tens** = 1 **hundred**.

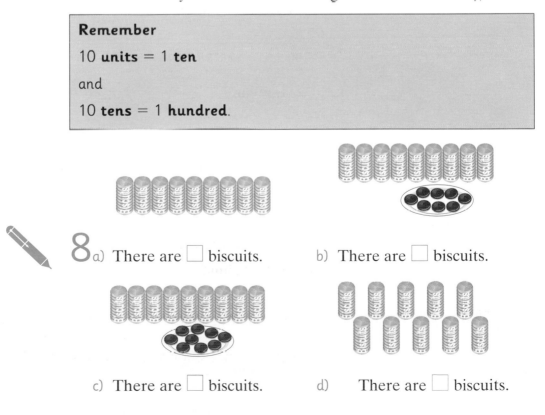

8 a) There are ☐ biscuits.   b) There are ☐ biscuits.

c) There are ☐ biscuits.   d)   There are ☐ biscuits.

9   Try Worksheet 1 *Draw and group again (1)*.

10   Rani wants to know how many badges Ella has collected.
He does not want to know the **exact** number.
He wants to know **roughly** – to the **nearest ten**.

Which ten are these numbers nearer to,   **9 tens**   or   **10 tens**?
                                          90   or   **100**?

Remember

Find out the middle number.

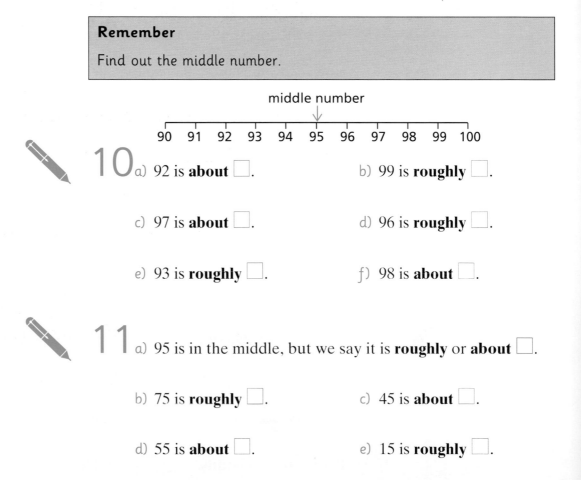

middle number

90   91   92   93   94   95   96   97   98   99   100

10 a) 92 is **about** ☐.   b) 99 is **roughly** ☐.

c) 97 is **about** ☐.   d) 96 is **roughly** ☐.

e) 93 is **roughly** ☐.   f) 98 is **about** ☐.

11 a) 95 is in the middle, but we say it is **roughly** or **about** ☐.

b) 75 is **roughly** ☐.   c) 45 is **about** ☐.

d) 55 is **about** ☐.   e) 15 is **roughly** ☐.

What do you notice about the **middle** numbers?

f) They all end in the digit ☐.

12   Ben's dad keeps 10 packets of biscuits in a large box.

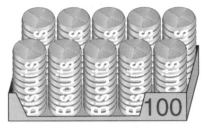

12 a) There are ☐ biscuits in a packet.

b) There are ☐ packets of biscuits in a box.

c) There are ☐ biscuits altogether in a box.

Which is worth more, one packet or one box?

d) One _____ is **worth more**.

13   How many biscuits are there altogether in each drawing?

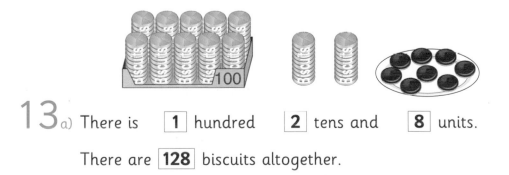

13 a) There is 1 hundred   2 tens and   8 units.

There are 128 biscuits altogether.

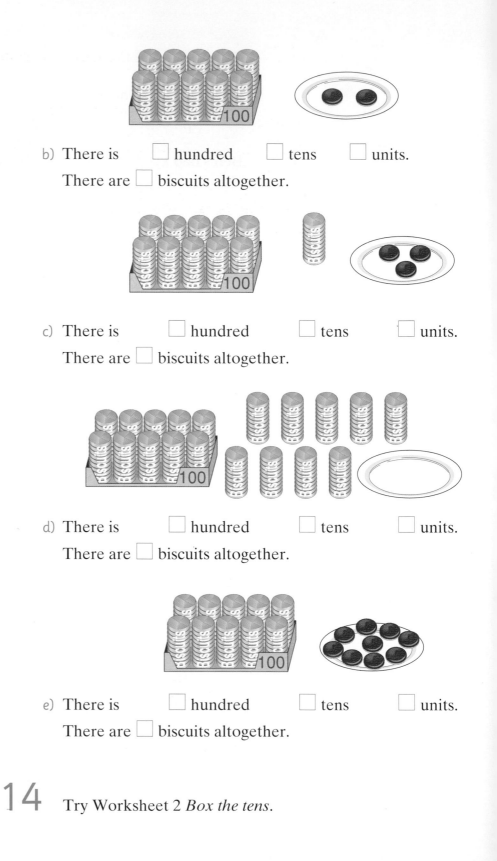

b) There is ☐ hundred ☐ tens ☐ units.
There are ☐ biscuits altogether.

c) There is ☐ hundred ☐ tens ☐ units.
There are ☐ biscuits altogether.

d) There is ☐ hundred ☐ tens ☐ units.
There are ☐ biscuits altogether.

e) There is ☐ hundred ☐ tens ☐ units.
There are ☐ biscuits altogether.

14 Try Worksheet 2 *Box the tens*.

15 Which is worth most, one (1p) coin, one (10p) coin or one (£1) coin?

> **Remember**
>
> Ten (1p) coins = one (10p) coin
>
> and
>
> ten (10p) coins = one (£1) coin.

15 a) One _____ coin is **worth most**.

b) One _____ coin has the **biggest value**.

Which is worth least, one (1p) coin, one (10p) coin or one (£1) coin?

c) One _____ coin is **worth least**.

d) One _____ coin has the **smallest value**.

16 How many biscuits are there altogether in each drawing?

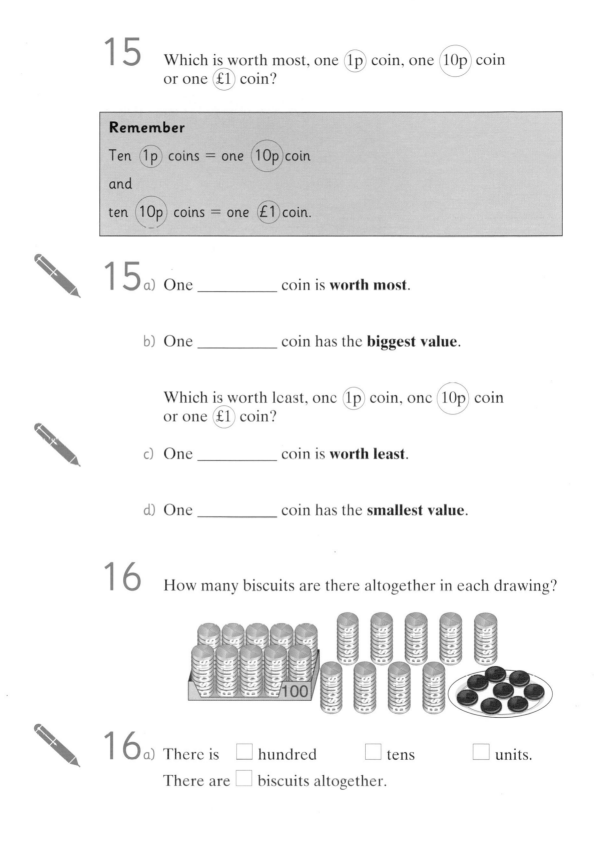

16 a) There is ☐ hundred ☐ tens ☐ units.

There are ☐ biscuits altogether.

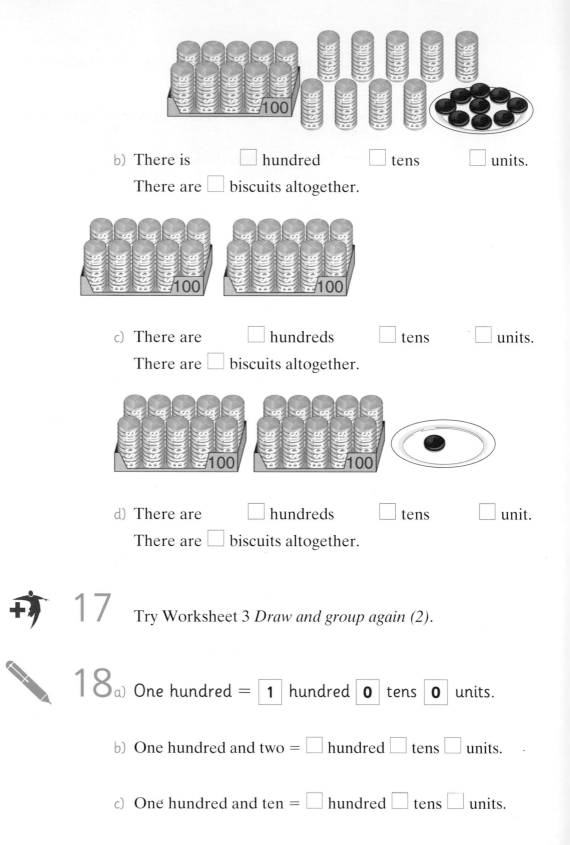

b) There is ☐ hundred ☐ tens ☐ units.
There are ☐ biscuits altogether.

c) There are ☐ hundreds ☐ tens ☐ units.
There are ☐ biscuits altogether.

d) There are ☐ hundreds ☐ tens ☐ unit.
There are ☐ biscuits altogether.

17 Try Worksheet 3 *Draw and group again (2).*

18 a) One hundred = ☐1☐ hundred ☐0☐ tens ☐0☐ units.

b) One hundred and two = ☐ hundred ☐ tens ☐ units.

c) One hundred and ten = ☐ hundred ☐ tens ☐ units.

d) One hundred and nineteen = ☐ hundred ☐ tens ☐ units.

e) One hundred and twenty = ☐ hundred ☐ tens ☐ units.

f) One hundred and ninety = ☐ hundred ☐ tens ☐ units.

g) One hundred and ninety-one = ☐ hundred ☐ tens ☐ units.

19 Try Worksheet 4 *Find the place*.

20 Copy these numbers and circle the digit that has the **biggest** value.

20 a) ① 3 1

b) 1 5 2

c) 1 4 1

d) 1 1 6

e) 1 0 0

f) 1 0 4

g) 1 1 8

h) 1 0 5

i) 1 1 3

j) 1 9 0

Say it, write it

21 Play a game of 'Say it, write it'.

> **Rules for 'Say it, write it'**
> 1 Get a friend to read the numbers in Question 20 aloud.
> 2 You write the numbers down.
> 3 Check to see how many you wrote correctly.
> 4 Change over. Now you read the numbers in Question 20 aloud.
> 5 Your friend writes the numbers down.
> 6 Check to see how many numbers your friend wrote correctly.
> 7 The winner is the player who wrote the most correct numbers.

 22 Try Worksheet 5 *What value?*

23 Draw these groups of lockers.
Fill in the number **before** and **after** each numbered locker.

10 50 79 99

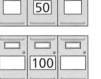
110 100 159 200

24 Two schools play cricket.

Season scores				
School A			**School B**	
Player 1	10 runs		Player 1	54 runs
Player 2	190 runs		Player 2	176 runs
Player 3	20 runs		Player 3	0 runs
Player 4	150 runs		Player 4	83 runs
Player 5	80 runs		Player 5	198 runs
Player 6	110 runs		Player 6	115 runs
Player 7	130 runs		Player 7	25 runs
Player 8	100 runs		Player 8	2 runs
Player 9	140 runs		Player 9	107 runs
Player 10	180 runs		Player 10	200 runs
Player 11	90 runs		Player 11	12 runs

24 a) Look at the **school A** team.
Put the runs in order, **biggest** number first.

b) Now do the same for **school B**.

c) You need

• a calculator.

Guess which school scored most runs altogether.
Now use your calculator to find the answer.

25 You need

• a calculator.

Take turns to put these numbers into the calculator.
Take turns to guess the answers.

Look at the calculator to see if you are right.

25

a)
0 + 10
20 + 10
50 + 10
70 + 10
30 + 10
60 + 10
10 + 10
40 + 10
80 + 10
90 + 10

b)
100 + 10
180 + 10
130 + 10
160 + 10
110 + 10
150 + 10
120 + 10
170 + 10
140 + 10
190 + 10

c)
121 + 1
121 + 10
129 + 1
100 + 1
90 + 1
189 + 10
189 + 1
9 + 1
99 + 1
199 + 1

 26 Try Worksheet Puzzle *Change the digit.*

27 You need

- Unit 3 Race against time cards Set 1 and Set 2
- your 'My maths record' sheet.

Race against time

1 Sort the Set 1 race cards – this side up. 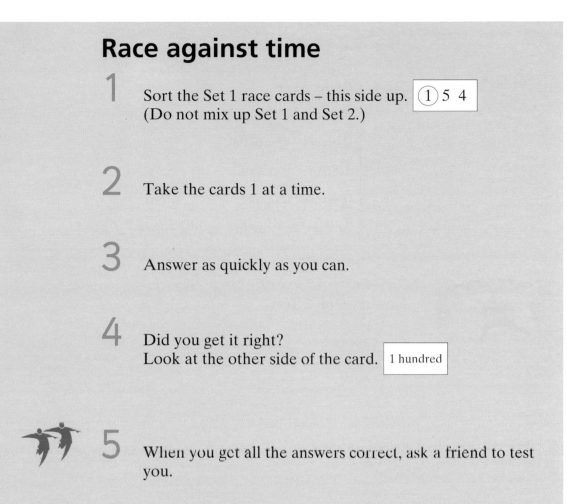 | (1) 5 4 |
(Do not mix up Set 1 and Set 2.)

2 Take the cards 1 at a time.

3 Answer as quickly as you can.

4 Did you get it right?
Look at the other side of the card. | 1 hundred |

 5 When you get all the answers correct, ask a friend to test you.

6 Now **Race against time**
Go for points!
Ask your teacher to test and time you.

7 Now try Set 2 – this side up. | to nearest ten 49 |

8   You can go for more points.

> **Remember**
>
> 3 errors – 1 point
>
> 2 errors – 2 points
>
> 1 error – 3 points
>
> 0 errors – 5 points
>
> Answer in 1 minute with 0 errors – 7 points

Now try Unit 3 Test.

Review 3

1 a) 14p **plus** 4p = ☐ p   b)   10p **plus** 8p = ☐ p

c) 9p **add** 8p = ☐ p   d)   11p **add** 9p = ☐ p

e) 5p **and** 8p = ☐ p   f)   5p **and** 9p – ☐ p

g) 9p + 9p = ☐ p   h)   7p + 7p = ☐ p

2 a) 5p **minus** 5p = ☐ p   b)   10p − 8p = ☐ p

c) 9p − 7p = ☐ p   d)   7p **minus** 3p = ☐ p

e) 10p − 1p = ☐ p   f)   2p − 1p = ☐ p

g) 8p **take away** 6p = ☐ p   h)   10p **take away** 4p = ☐ p

3 Write in order, **smallest** number first.

   13   88   16   33   51   7

4   Which angles are **right angles**?

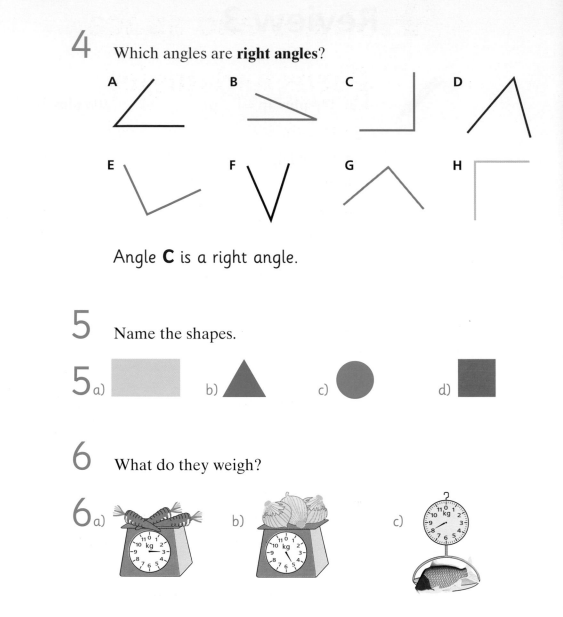

A   B   C   D

E   F   G   H

Angle **C** is a right angle.

5   Name the shapes.

5 a)   b)   c)   d)

6   What do they weigh?

6 a)   b)   c)

4 Shape and space

Turns and angles

Unit 4 words

quarter	half	turn
angle	hexagon	pentagon
greater than	shape	clockwise
right angle	about	less than

Remember

Examples are shown in red.

 means copy and complete.

 You need

- a set of Unit 4 vocabulary Snap cards.

 Play a game of Snap to help you learn the words.

Try the **word test** to get some points.

Clockwise turns

The hands on a clock face turn **clockwise**.

start

Remember

Clockwise is the **same** way as a clock.

The hands on these clock faces have turned:
one quarter turn clockwise   **two quarter** turns clockwise
(or **half** a turn clockwise)

three quarter turns clockwise   **four quarter** turns clockwise
(or a **full** turn clockwise)

1 How many **quarter** turns **clockwise** has the hand on each clock turned?

1 a)    a) **one quarter** turn clockwise

Anticlockwise turns

We call turns made in the **opposite** direction to a clock **anticlockwise** turns.

Remember
Anticlockwise is the **opposite** way to a clock.

anticlockwise – opposite to a clock   clockwise – same as a clock

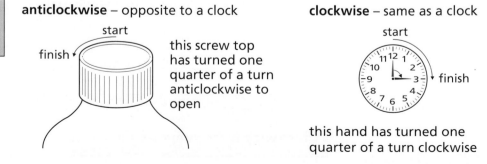

this screw top has turned one quarter of a turn anticlockwise to open

this hand has turned one quarter of a turn clockwise

2 How many **quarter** turns **anticlockwise** have these tops turned through?

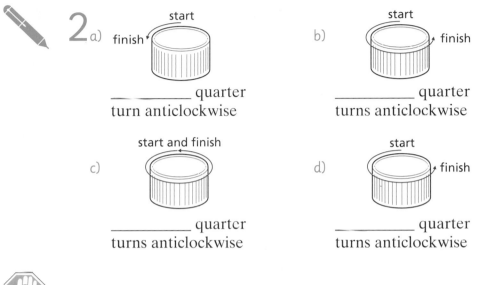

a) _____ quarter turn anticlockwise

b) _____ quarter turns anticlockwise

c) _____ quarter turns anticlockwise

d) _____ quarter turns anticlockwise

e) Try to open some different bottle tops.

Which way do they turn most often, **clockwise** or **anticlockwise**?

3 Guess if these turns are clockwise or anticlockwise.

Get a partner to do the turns to help you.

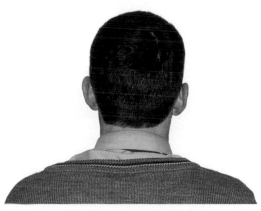

start

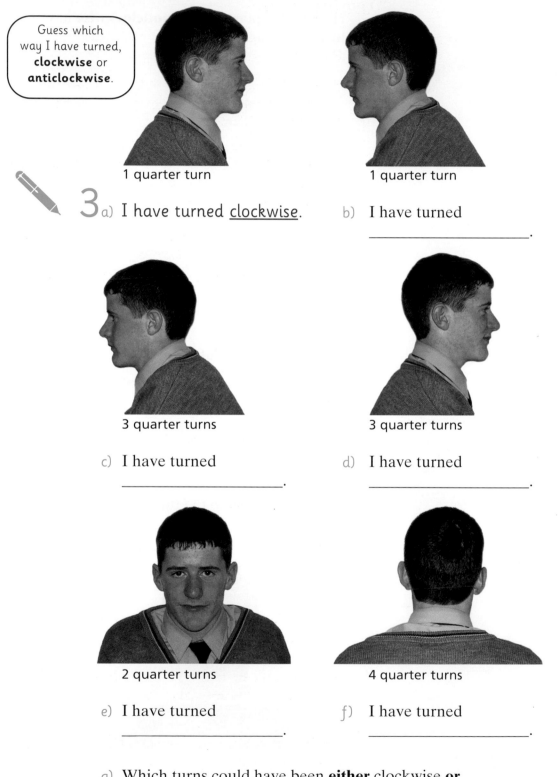

Guess which way I have turned, **clockwise** or **anticlockwise**.

1 quarter turn

1 quarter turn

3 a) I have turned <u>clockwise</u>.

b) I have turned

_____.

3 quarter turns

3 quarter turns

c) I have turned

_____.

d) I have turned

_____.

2 quarter turns

4 quarter turns

e) I have turned

_____.

f) I have turned

_____.

g) Which turns could have been **either** clockwise **or** anticlockwise?

Turning game

4a) Play the 'Turning game'.

Always start with your back to your partner.

> **Rules for the 'Turning game'**
>
> **1** Your partner covers their eyes or turns away.
> **2** You turn clockwise or anticlockwise to a new position. Count your quarter turns as you do so.
> **3** When you are in your new position, tell your partner to look.
> **4** Tell your partner the number of quarter turns you have made.
> **5** Your partner guesses the direction, **clockwise** or **anticlockwise** or **either**.
> **6** Your partner then makes some turns and you guess the direction.

b) We can also call **one quarter turn**
a _____ turn.

Angles

An **angle** is the amount something turns.

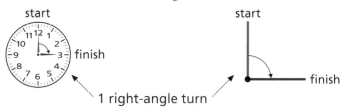

1 right-angle turn

5 You need

- two pieces of Meccano, or two Geostrips, or two strips of card.

5a) Make an **angle maker** from two pieces of Meccano, or two Geostrips, or two strips of card.

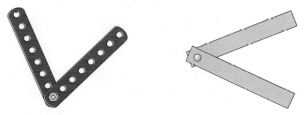

b) Open one arm of your
angle maker to make
different angles.
The **more** you open the
arms of your angle maker,
the **bigger** the angle.

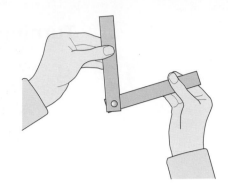

 **6** a) Use your angle maker to
make two angles, a **narrow** angle and a **wide** angle.

b) Draw the angles in your book. Label the angles **narrow** and
wide.

c) List these angles in order, **widest** angle first:

 7 Try Worksheet 1 *Sticky angles*.

Remember

We can call a **quarter turn** a **right angle**.

 8 You need

• a piece of scrap paper.

8 a) Make a **right angle checker** from a piece of scrap paper,
like this.

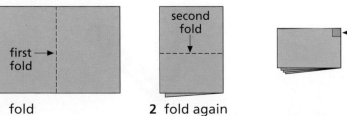

1 fold   **2** fold again

b) Use your right angle checker to find out whether these angles are:

greater than a right angle
less than a right angle
or **about** a right angle.

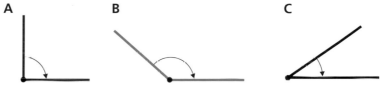

 9   Try Worksheet 2 *Mixed angles*.

10   Look at these turning lines.

1 right angle clockwise   **2** right angles clockwise (or a **straight turn**)   **3** right angles clockwise   **4** right angles clockwise (or a **full turn** clockwise)

How many **right angles clockwise** have these lines turned?

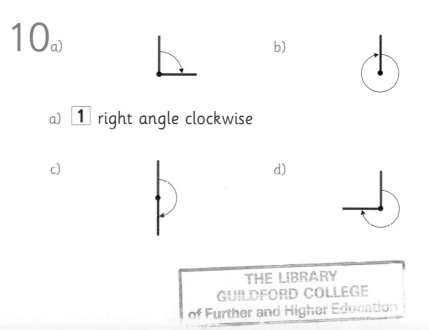

10a)

b)

a) ☐**1** right angle clockwise

c)

d)

11 How many **right angles anticlockwise** have these lines turned?

11 a) 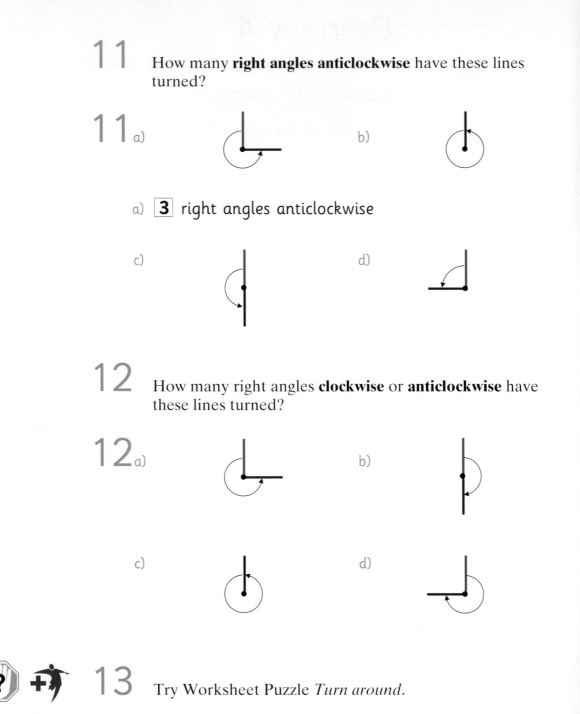 b)

a) **3** right angles anticlockwise

c)   d)

12 How many right angles **clockwise** or **anticlockwise** have these lines turned?

12 a)   b)

c)   d)

 **13** Try Worksheet Puzzle *Turn around.*

Now try Unit 4 Test.

Review 4

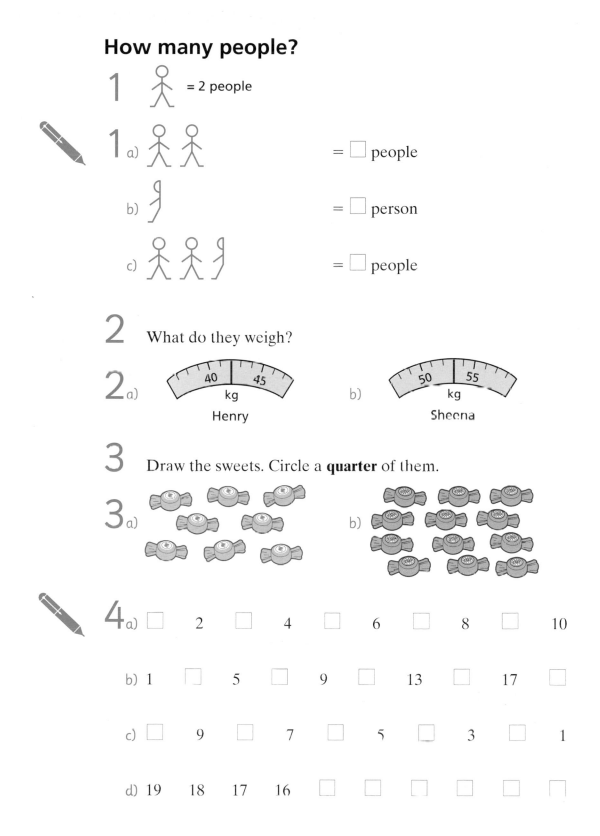

How many people?

1 = 2 people

1 a) = ☐ people

b) = ☐ person

c) = ☐ people

2 What do they weigh?

2 a) 40 | 45
kg
Henry

b) 50 | 55
kg
Sheena

3 Draw the sweets. Circle a **quarter** of them.

3 a)

b)

4 a) ☐ 2 ☐ 4 ☐ 6 ☐ 8 ☐ 10

b) 1 ☐ 5 ☐ 9 ☐ 13 ☐ 17 ☐

c) ☐ 9 ☐ 7 ☐ 5 ☐ 3 ☐ 1

d) 19 18 17 16 ☐ ☐ ☐ ☐ ☐ ☐

5 Write down these numbers. Circle the **odd** numbers.

1   2   3   4   5   6   7   8   9   10

6 Write down the answers to each sum.

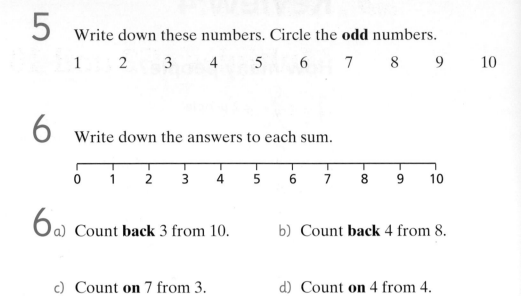

0   1   2   3   4   5   6   7   8   9   10

6 a) Count **back** 3 from 10.   b) Count **back** 4 from 8.

c) Count **on** 7 from 3.   d) Count **on** 4 from 4.

5 : Number

Multiples of 2 and 10

Unit 5 words

makes	multiply	times
worth	value	middle number
hundred	single	clockwise
right angle	tens	units

Remember

Examples are shown in red.

 means copy and complete.

 You need

- a set of Unit 5 vocabulary Snap cards.

 Play a game of Snap to help you learn the words.

 Try the **word test** to get some points.

1 a) One bicycle has ☐ wheels. b) Two bicycles have ☐ wheels.

c) Three bicycles have ☐ wheels.

d) Four bicycles have ☐ wheels.

 2 Try Worksheet 1 *Pairs.*

3 Draw the groups of buttons.

How many holes has each group of buttons?

3 a)

b)

b) There are **4** holes altogether.

c) ●● ●● ●●

d) ●● ●● ●● ●●

e) ●● ●● ●● ●● ●●

f) ●● ●● ●● ●● ●● ●●

g) ●● ●● ●● ●● ●● ●●

h) ●● ●● ●● ●● ●● ●● ●●

i) ●● ●● ●● ●● ●● ●● ●●

j) ●● ●● ●● ●● ●● ●● ●● ●●

4 Try Worksheet 2 *How many times? (1)*.

5 You need

- squared paper.

Look at the **rows of 2**. Copy them on to squared paper.

5 a) 1 × 2   b) 2 × 2   c) 3 × 2   d) 4 × 2   e) 5 × 2

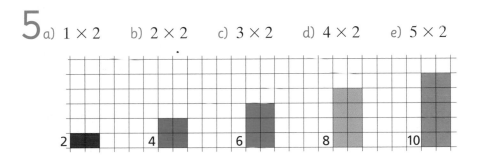

Continue drawing the patterns of squares for:

f) 6 × 2   g) 7 × 2   h) 8 × 2   i) 9 × 2   j) 10 × 2

Jumping in twos

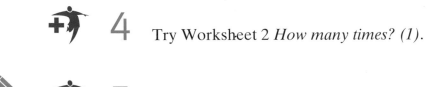

6 a) 2 jumps of two is 2 × ☐.

b) 2 × 2 = ☐

c) Try Worksheet 3 *Jumps of two*.

7   (You can use the number line in **Question 6** to help you.)

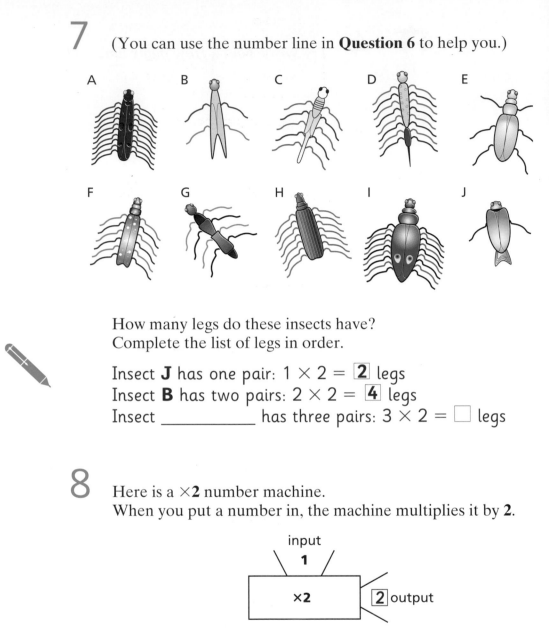

How many legs do these insects have?
Complete the list of legs in order.

Insect **J** has one pair: $1 \times 2 = $ [2] legs
Insect **B** has two pairs: $2 \times 2 = $ [4] legs
Insect _____ has three pairs: $3 \times 2 = $ ☐ legs

8   Here is a ×**2** number machine.
When you put a number in, the machine multiplies it by **2**.

input
1

| ×**2** | [2] output |

Draw the machines and complete the sentences.

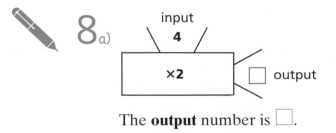

8 a)

input
4

| ×**2** | ☐ output |

The **output** number is ☐.

b)

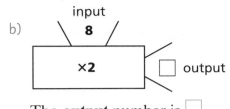

input
8
×2
☐ output

The **output** number is ☐.

9   Try Worksheet 4 *Number machines (1).*

10 a) ☐ × 2 = 6   b) ☐ × 2 = 12

   a) **3** × 2 = 6

   c) ☐ × 2 = 2   d) ☐ × 2 = 18

   e) ☐ × 2 = 10   f) ☐ × 2 = 16

   g) ☐ × 2 = 4   h) ☐ × 2 = 14

   i) ☐ × 2 = 8   j) ☐ × 2 = 20

11   Look at the dot patterns.

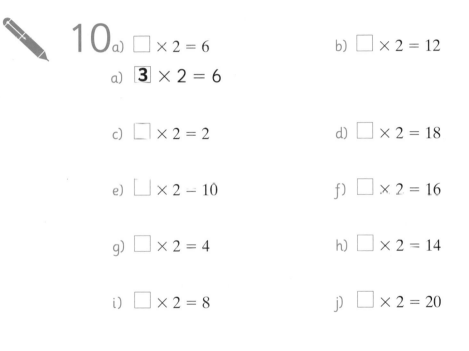

3 rows of 2 make 6

2 rows of 3 make 6

3 rows of 2 and **2 rows of 3** give the **same** answer.

Copy and complete the **multiple pairs**:

11 a) 2 × 3 = ☐   3 × 2 = ☐

Draw the patterns of dots and make your own multiple pairs for these numbers:

b) 8   c) 10   d) 12

12 Multiply to find the answers.

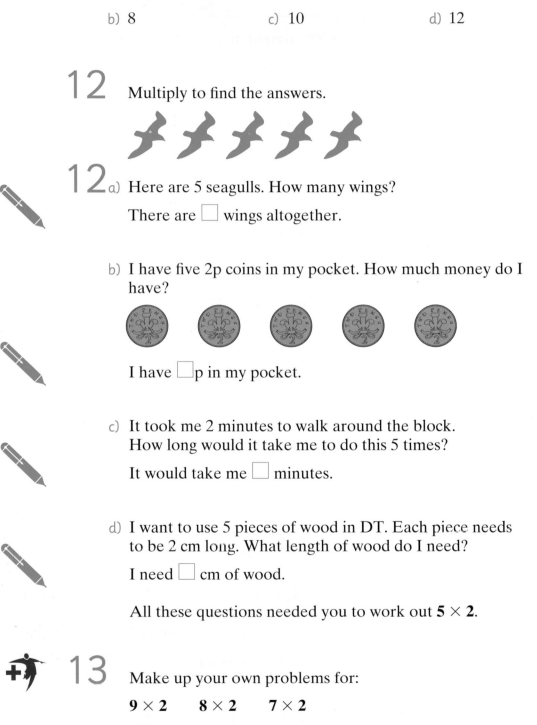

12 a) Here are 5 seagulls. How many wings?

There are ☐ wings altogether.

b) I have five 2p coins in my pocket. How much money do I have?

I have ☐ p in my pocket.

c) It took me 2 minutes to walk around the block. How long would it take me to do this 5 times?

It would take me ☐ minutes.

d) I want to use 5 pieces of wood in DT. Each piece needs to be 2 cm long. What length of wood do I need?

I need ☐ cm of wood.

All these questions needed you to work out **5 × 2**.

13 Make up your own problems for:

9 × 2   8 × 2   7 × 2

Make a display.

14   There are 10 eggs in each box.
      How many tens are there altogether?

14 a) There are **10** eggs in **1** box.

   There is **1** ten.

   b) There are ☐ eggs in **2** boxes.
      There are ☐ tens.

   c) There are ☐ eggs in **3** boxes.
      There are ☐ tens.

   d) There are ☐ eggs
      in **4** boxes.
      There are ☐ tens.

15   There are 10 pencils in each bundle.
      How many pencils are there altogether?

15 a)   b)

a) **5** groups of 10 pencils   ☐ groups of 10 pencils
    5 tens   ☐ tens
    50 pencils altogether   ☐ pencils altogether

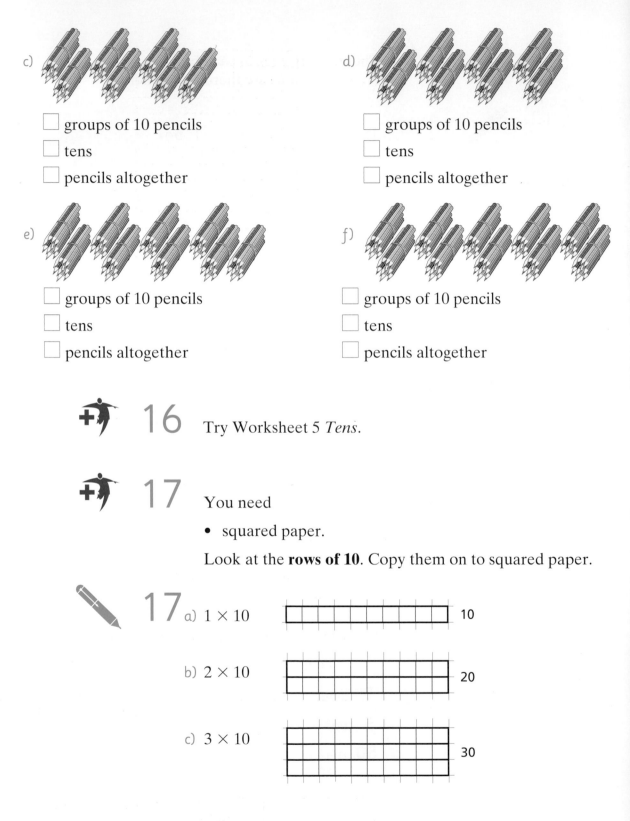

c)
☐ groups of 10 pencils
☐ tens
☐ pencils altogether

d)
☐ groups of 10 pencils
☐ tens
☐ pencils altogether

e)
☐ groups of 10 pencils
☐ tens
☐ pencils altogether

f)
☐ groups of 10 pencils
☐ tens
☐ pencils altogether

16 Try Worksheet 5 *Tens*.

17 You need

• squared paper.

Look at the **rows of 10**. Copy them on to squared paper.

17 a) 1×10 10

b) 2×10 20

c) 3×10 30

d) 4 × 10

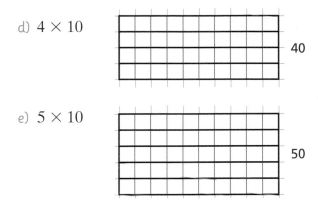

40

e) 5 × 10

50

Continue drawing the patterns of squares for:

f) 6 × 10   g) 7 × 10   h) 8 × 10

i) 9 × 10   j) 10 × 10

18 Try Worksheet 6 *How many times? (2)*.

19 Here is a ×**10** number machine.

input
☐
×**10**   ☐ output

Draw the machines and complete the sentences.

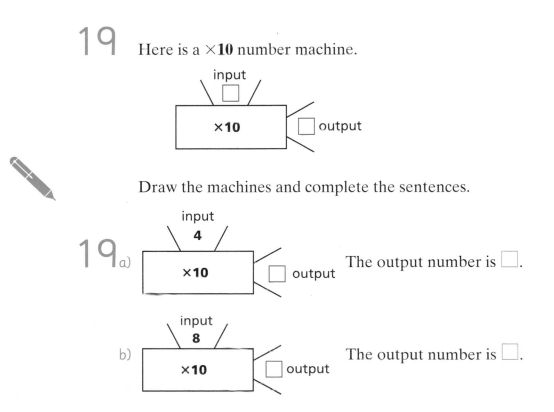

19 a)

input
4
×**10**   ☐ output   The output number is ☐.

b)

input
8
×**10**   ☐ output   The output number is ☐.

20 Try Worksheet 7 *Number machines (2)*.

21 a) **3** × 10 = 30

b) ☐ × 10 = 60   c) ☐ × 10 = 10   d) ☐ × 10 = 90

e) ☐ × 10 = 50   f) ☐ × 10 = 80   g) ☐ × 10 = 20

h) ☐ × 10 = 70   i) ☐ × 10 = 40   j) ☐ × 10 = 100

22 Look at the dot patterns.

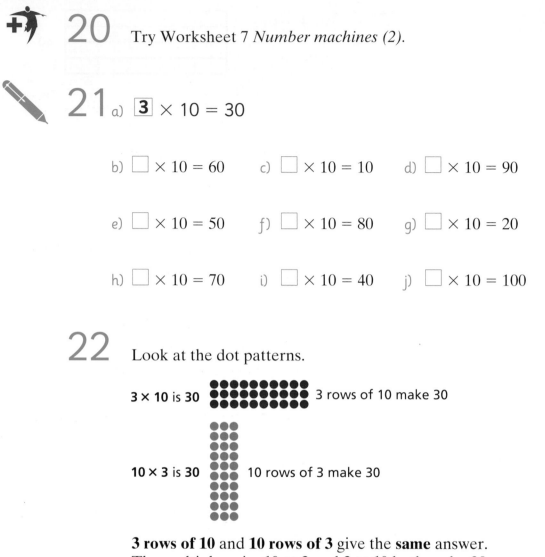

3 × 10 is 30   3 rows of 10 make 30

10 × 3 is 30   10 rows of 3 make 30

3 rows of 10 and **10 rows of 3** give the **same** answer.
The multiple pairs **10 × 3** and **3 × 10** both make 30.

Make your own **multiple pairs** for these numbers:

22 a) 10   b) 20   c) 40   d) 50

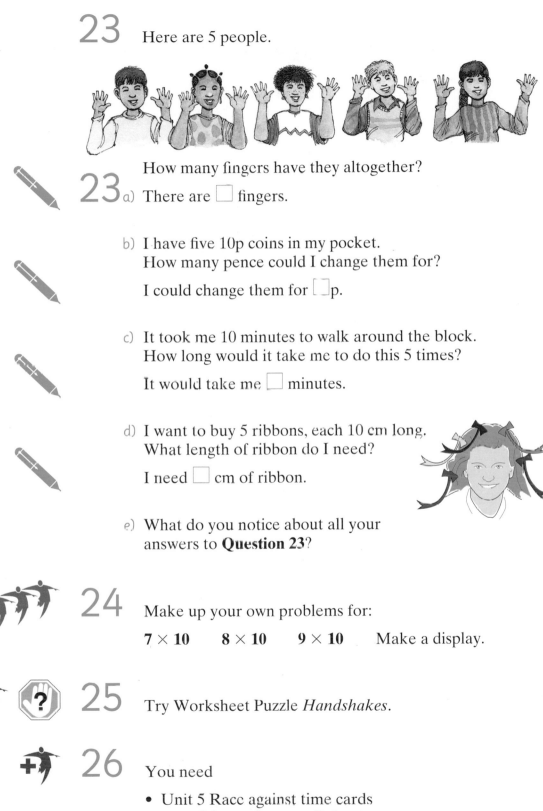

23 Here are 5 people.

How many fingers have they altogether?

23 a) There are ☐ fingers.

b) I have five 10p coins in my pocket.
How many pence could I change them for?

I could change them for ☐ p.

c) It took me 10 minutes to walk around the block.
How long would it take me to do this 5 times?

It would take me ☐ minutes.

d) I want to buy 5 ribbons, each 10 cm long.
What length of ribbon do I need?

I need ☐ cm of ribbon.

e) What do you notice about all your
answers to **Question 23**?

24 Make up your own problems for:

7 × 10 **8 × 10** **9 × 10** Make a display.

25 Try Worksheet Puzzle *Handshakes*.

26 You need

- Unit 5 Race against time cards
- your 'My maths record' sheet.

Race against time

1 Sort the race cards – this side up $\boxed{2 \times 10}$

2 Take the cards 1 at a time.

3 Answer as quickly as you can.

4 Did you get it right?

Look at the other side of the card. $\boxed{20}$

5 When you get all the answers correct, ask a friend to test you.

6 Now **Race against time**

Go for points!
Ask your teacher to test and time you.

Remember

3 errors – 1 point

2 errors – 2 points

1 error – 3 points

0 errors – 5 points

Answer in 1 minute with 0 errors – 7 points

Now try Unit 5 Test.

Review 5

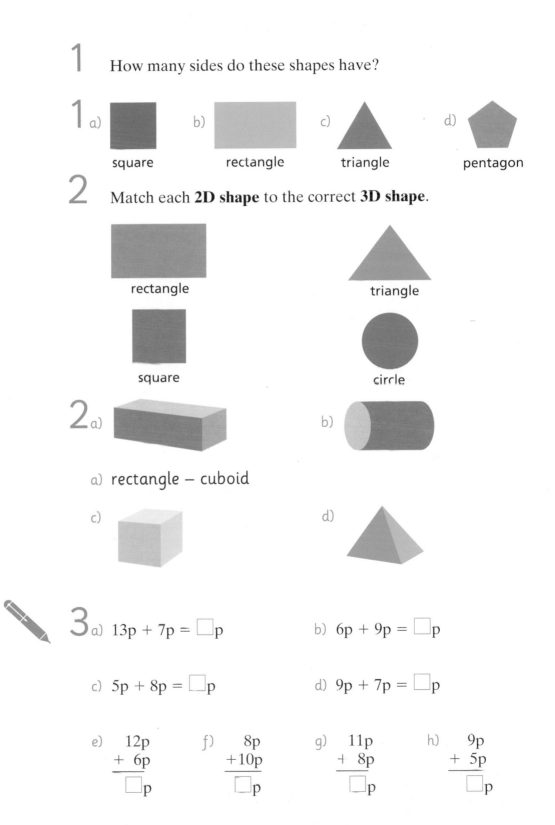

1 How many sides do these shapes have?

a) square

b) rectangle

c) triangle

d) pentagon

2 Match each **2D shape** to the correct **3D shape**.

rectangle

triangle

square

circle

a)

b)

a) rectangle – cuboid

c)

d)

3 a) $13p + 7p = \boxed{}p$ b) $6p + 9p = \boxed{}p$

c) $5p + 8p = \boxed{}p$ d) $9p + 7p = \boxed{}p$

e) $\begin{array}{r} 12p \\ +\ 6p \\ \hline \boxed{}p \end{array}$ f) $\begin{array}{r} 8p \\ +10p \\ \hline \boxed{}p \end{array}$ g) $\begin{array}{r} 11p \\ +\ 8p \\ \hline \boxed{}p \end{array}$ h) $\begin{array}{r} 9p \\ +\ 5p \\ \hline \boxed{}p \end{array}$

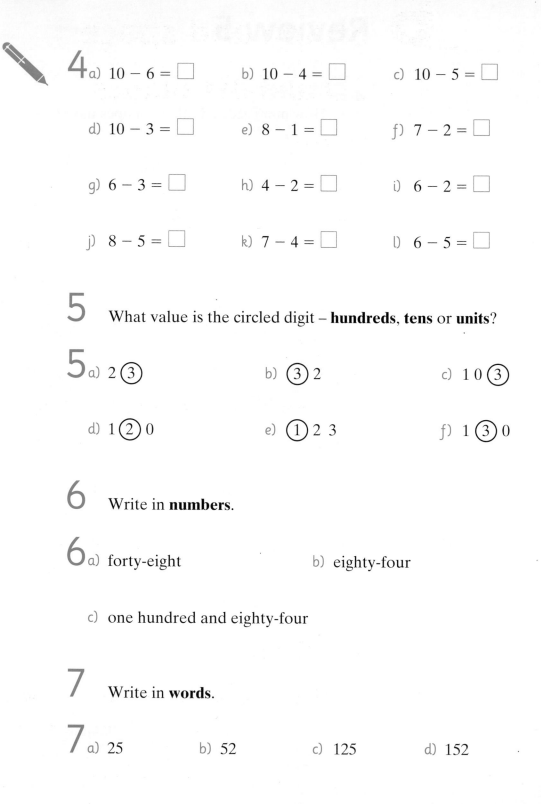

4 a) 10 − 6 = ☐   b) 10 − 4 = ☐   c) 10 − 5 = ☐

d) 10 − 3 = ☐   e) 8 − 1 = ☐   f) 7 − 2 = ☐

g) 6 − 3 = ☐   h) 4 − 2 = ☐   i) 6 − 2 = ☐

j) 8 − 5 = ☐   k) 7 − 4 = ☐   l) 6 − 5 = ☐

5 What value is the circled digit – **hundreds**, **tens** or **units**?

5 a) 2 ③   b) ③ 2   c) 1 0 ③

d) 1 ② 0   e) ① 2 3   f) 1 ③ 0

6 Write in **numbers**.

6 a) forty-eight   b) eighty-four

c) one hundred and eighty-four

7 Write in **words**.

7 a) 25   b) 52   c) 125   d) 152

6 Shape and space

2D and 3D shapes

Unit 6 words

rectangle	triangle	curved
straight	cube	cuboid
cylinder	sides	edges
faces	symmetry	pyramid

Remember

Examples are shown in red.

 means copy and complete.

 You need

- a set of Unit 6 vocabulary Snap cards.

 Play a game of Snap to help you learn the words.

 Try the **word test** to get some points.

A 2D shapes

1   Name the shape. Match it to the correct group.

 A   group W

 B   group X

C   group Y

D   group Z

1 a) Shape **A** is a _____.

It matches group _____.

b) Shape **B** is a _____.

It matches group _____.

c) Shape **C** is a _____.

It matches group _____.

d) Shape **D** is a _____.
It matches group _____.

Name the shape. Match it to the correct group.

E   group **G**

F   group **H**

e) Shape **E** is a _____.

It matches group _____.

f) Shape **F** is a _____.

It matches group _____.

g) Explain why shape **E** can match **two** groups and
shape **F** can match only **one**.

Write your answer down or tell your teacher.

Draw my shape

2 You need

- a sheet of squared spotty paper.

Rules for 'Draw my shape'

Player A
1 Look at one of the shapes in **Question 1**.
2 Pick one of them, but do not say which one.
3 Tell your partner how to draw the shape you are thinking of, using the words below. Take as many turns as you need.

Words to help you tell your partner how to draw the shape:

draw a straight line   going up
draw a right angle turn   going down
draw a curved line   going left
opposite sides   going right
the same length   join to starting point

Player B
1 Draw the lines your partner tells you to draw on spotty paper.
2 What shape have you drawn?

---

2 a) Play a game of 'Draw my shape'.

b) Did Player B draw the right shape?

Does it look like the shape in the book?

Could it look better? How?

c) Take turns to draw and guess the other shapes.

3 a) How many different shapes can you see in this flag?

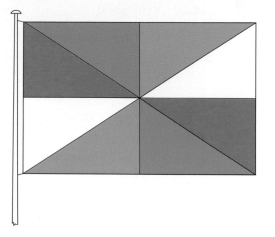

b) Draw the flag and colour it in a different way to show different shapes.

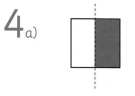

c) Try Worksheet 1 *Find the shapes*.

4 Put a mirror on the dotted line ------------ .

Write the name of the shape you see.

4 a)

It is a **square**.

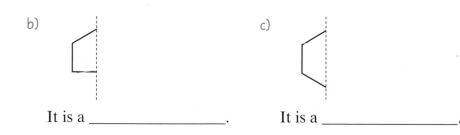

b)

It is a _____.

c)

It is a _____.

> **Remember**
>
> We call these lines **lines of symmetry**.
>
> We call shapes that fold exactly **symmetrical shapes**.

 5 Try Worksheet 2 *Cut and fold (1)* and
Worksheet 3 *Colour the leaves*.

6 If you folded on the dotted line --------------, would you fold
along the line of symmetry of these shapes?
Answer **yes** or **no**.

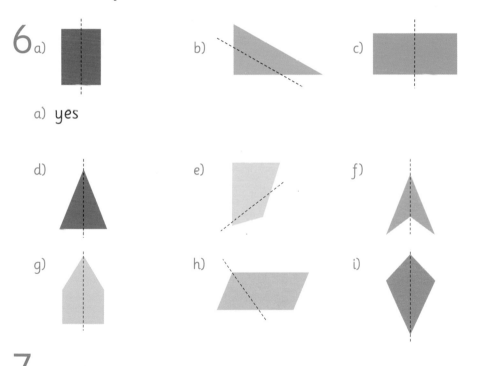

6 a)   b)   c)

a) yes

d)   e)   f)

g)   h)   i)

 7 Try Worksheet 4 *Cut and fold (2)*.

8 How many different lines of symmetry does a **square** have?
Use Worksheet 5 *Square fold* to help you find out.

9 How many lines of symmetry do these shapes have?

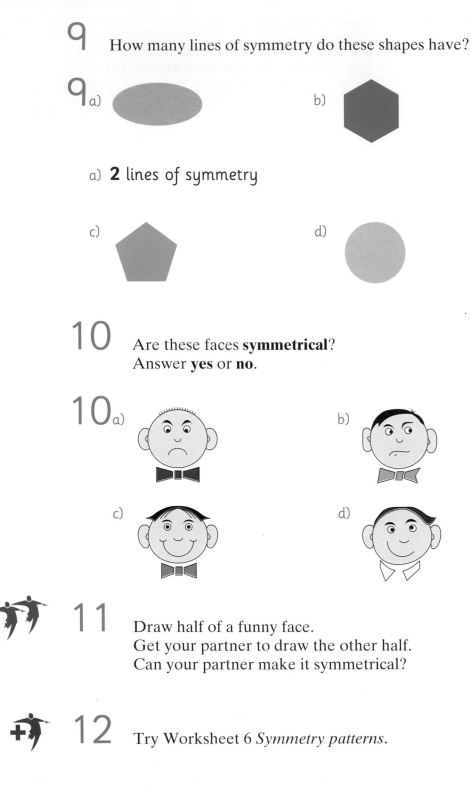

9 a) b)

a) **2** lines of symmetry

c) d)

10 Are these faces **symmetrical**?
Answer **yes** or **no**.

10 a) b)

c) d)

11 Draw half of a funny face.
Get your partner to draw the other half.
Can your partner make it symmetrical?

12 Try Worksheet 6 *Symmetry patterns*.

13 Make two lists to sort the shapes.
List the shapes that are **symmetrical**.
List the shapes that are **not symmetrical**.

Remember

Symmetrical shapes will fold into 2 equal parts.

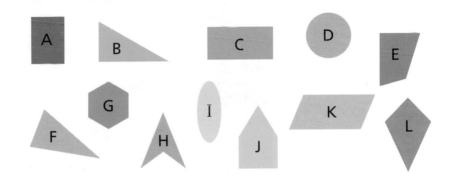

Symmetrical	Not symmetrical
A	B

14 To make a telephone call, a phone card must go into the slot the right way round.

How many ways **could** it fit into the slot?
You can turn the card over.

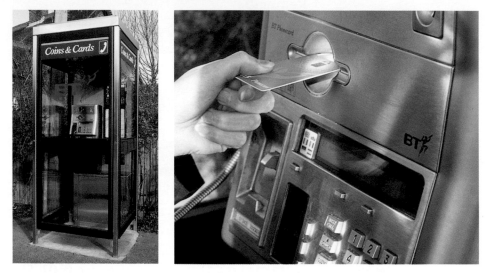

Now try Unit 6A Test.

B 3D shapes

1 Name the shape. Match it to the correct group.

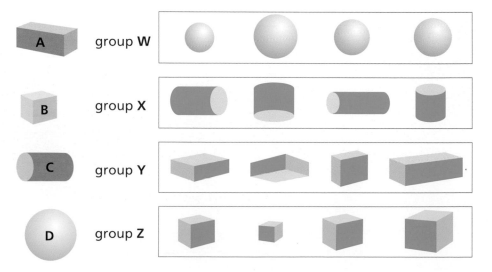

A   group **W**

B   group **X**

C   group **Y**

D   group **Z**

1 a) Shape **A** is a _____.

It matches group _____.

b) Shape **B** is a _____.

It matches group _____.

c) Shape **C** is a _____.

It matches group _____.

d) Shape **D** is a _____.

It matches group _____.

2 You need

- 3D shapes.

These are pyramids in Egypt.

Find the pyramid with a **square base** and the pyramid with a **triangle base**.

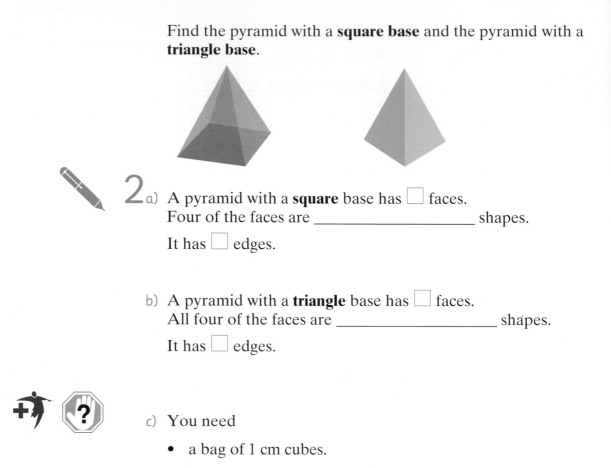

2 a) A pyramid with a **square** base has ☐ faces.
Four of the faces are _____ shapes.
It has ☐ edges.

b) A pyramid with a **triangle** base has ☐ faces.
All four of the faces are _____ shapes.
It has ☐ edges.

c) You need

• a bag of 1 cm cubes.

What kind of base do you think the pyramids in Egypt had, **square** or **triangle**?
Use cubes to build a small pyramid of your own.

Guess the shape

3 a)

Shape A
It is a 3D shape.
It has six faces.
All the faces are flat.
All the faces are square.
It **can't** roll easily.
It is a _____.

b)

Shape B
It is a 3D shape.
It has six faces.
All the faces are flat.
Four faces are rectangles.
It **can't** roll easily.
It is a _____.

c)

Shape C
It is a 3D shape.
It has three faces.
One face is curved.
Two faces are flat.
It **can** roll easily.
It is a _____.

4 The solid **shapes** were cut and used to make the **prints**.
Match the prints to the shapes.

shapes

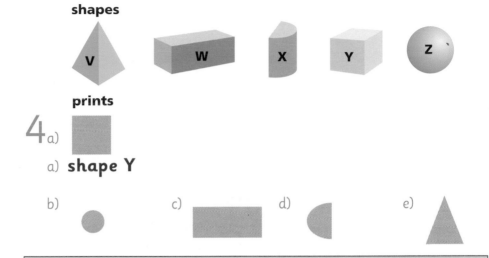

prints

4 a)

a) **shape Y**

b) c) d) e)

Remember

If something can be cut so that the 2 parts are exactly the same, we say that it is **symmetrical**.

---

5 A line is drawn where these fruit will be cut.
Do the cuts show that the fruit is **symmetrical**?
Answer **yes** or **no**.

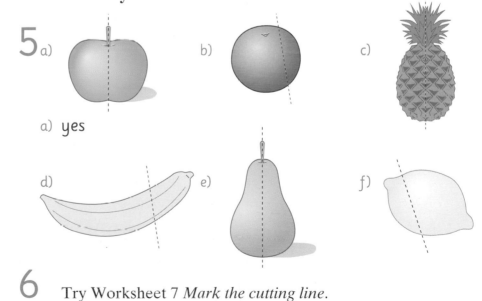

5 a)

b)

c)

a) **yes**

d)

e)

f)

6 Try Worksheet 7 *Mark the cutting line*.

7 Write down **two** ways these shapes are alike.
You can use some of these ideas to help you.

2D shape	or	3D shape
has same number of sides?		has same number of faces?
has all sides equal?		has flat faces?
is symmetrical?		has curved faces?
has four right angles?		**can** roll easily?
		can't roll easily?

7 a)

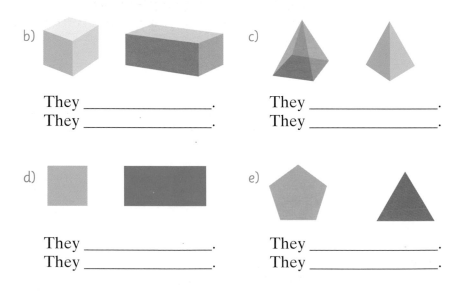

They **are 3D shapes**.
They **can roll easily**.
Can you think of any other ways they are alike?

b)

They _____.
They _____.

c)

They _____.
They _____.

d)

They _____.
They _____.

e)

They _____.
They _____.

8 You need

- these 2D shapes

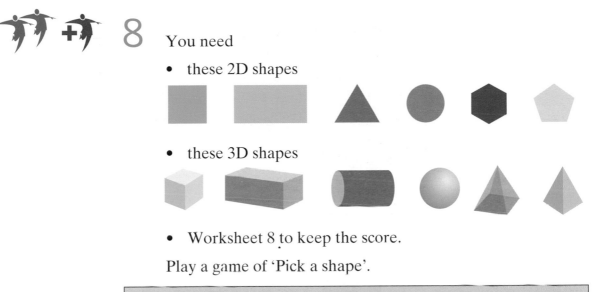

- these 3D shapes

- Worksheet 8 to keep the score.

Play a game of 'Pick a shape'.

Rules for 'Pick a shape'

1 Player A thinks of **two** ways that two shapes are the same.
2 Player A picks the shapes out and shows them to player B.
3 Player B guesses two ways the shapes are the same.
4 Take turns to pick and guess.
5 You could go on to try guessing **three** ways that two shapes are the same.

---

9 Find different shapes around your school or on your way to school. Use Worksheet 9 *Shape trail*.

10 Try Worksheet Puzzle *Flags*.

Now try Unit 6B Test.

Review 6

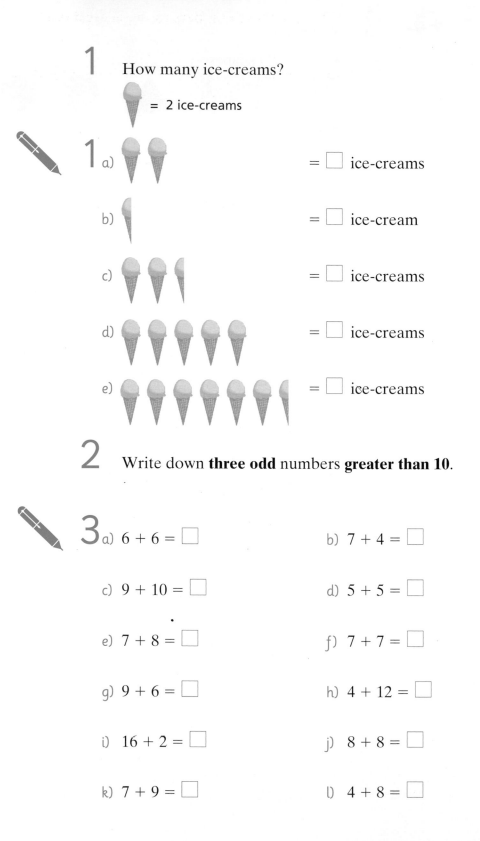

1 How many ice-creams?

= 2 ice-creams

1 a) = ☐ ice-creams

b) = ☐ ice-cream

c) = ☐ ice-creams

d) = ☐ ice-creams

e) = ☐ ice-creams

2 Write down **three odd** numbers **greater than 10**.

3 a) 6 + 6 = ☐   b) 7 + 4 = ☐

c) 9 + 10 = ☐   d) 5 + 5 = ☐

e) 7 + 8 = ☐   f) 7 + 7 = ☐

g) 9 + 6 = ☐   h) 4 + 12 = ☐

i) 16 + 2 = ☐   j) 8 + 8 = ☐

k) 7 + 9 = ☐   l) 4 + 8 = ☐

4   What are these numbers to the **nearest ten**?

4 a) 43 is roughly ☐.   b) 94 is roughly ☐.

c) 95 is roughly ☐.

5   How many **right angles** in each shape?

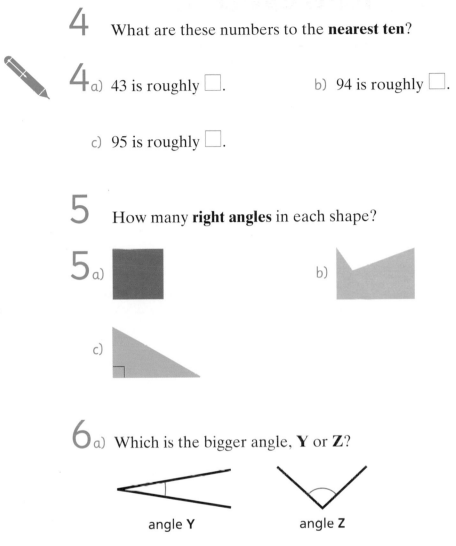

5 a)

b)

c)

6 a) Which is the bigger angle, **Y** or **Z**?

angle **Y**   angle **Z**

b) Which angle is on the **left**, **Y** or **Z**?

7 Handling data

Block graphs

Unit 7 words

sphere	circle	square
fifty	sixty	seventy
pictogram	block graph	tally
most	fewer	pyramid

Remember

Examples are shown in red.

means copy and complete.

 You need

- a set of Unit 7 vocabulary Snap cards.

 Play a game of Snap to help you learn the words.

 Try the **word test** to get some points.

1 This is the graph you saw in Unit 2, page 20.
Pupils in tutor group 9B found out what time the people in their tutor group **go to bed**.

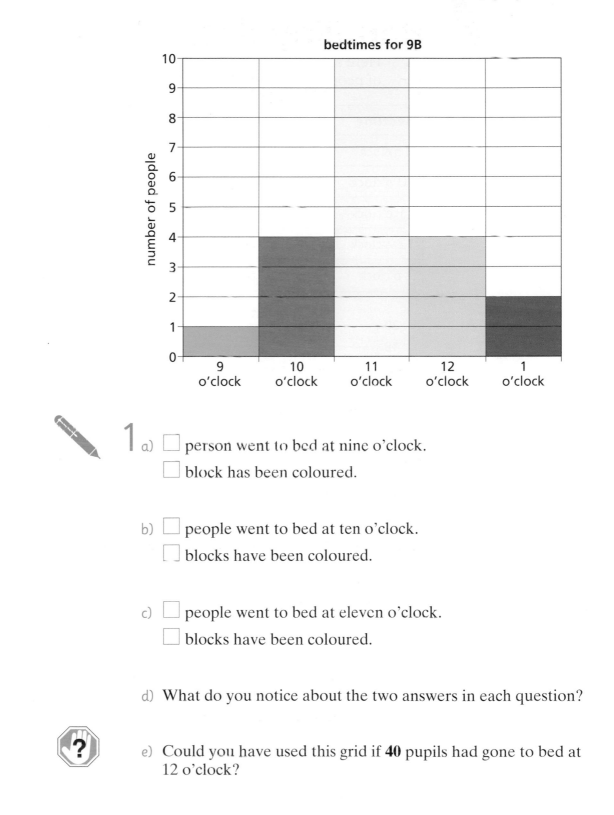

1 a) ☐ person went to bed at nine o'clock.

☐ block has been coloured.

b) ☐ people went to bed at ten o'clock.

☐ blocks have been coloured.

c) ☐ people went to bed at eleven o'clock.

☐ blocks have been coloured.

d) What do you notice about the two answers in each question?

e) Could you have used this grid if **40** pupils had gone to bed at 12 o'clock?

2 Here are the results of a survey done with a **different** group of people.

Bedtime	Tally marks
9 o'clock	II
10 o'clock	IIII
11 o'clock	HHI III
12 o'clock	HHI HHI
1 o'clock	HHI HHI II

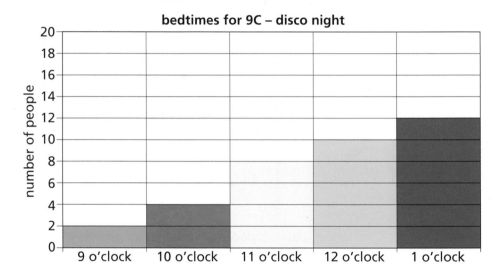

bedtimes for 9C – disco night

Look at the tally table and block graph.

2 a) ☐2 went to bed at nine o'clock.

☐1 block is coloured.

b) ☐ went to bed at ten o'clock.

☐ blocks are coloured.

c) ☐ went to bed at eleven o'clock.

☐ blocks are coloured.

d) ☐ went to bed at twelve o'clock.

☐ blocks are coloured.

e) ☐ went to bed at one o'clock.

☐ blocks are coloured.

f) What do you notice about the two answers in each question?

g) What is the largest number of blocks coloured?

3 How many people do these columns stand for?

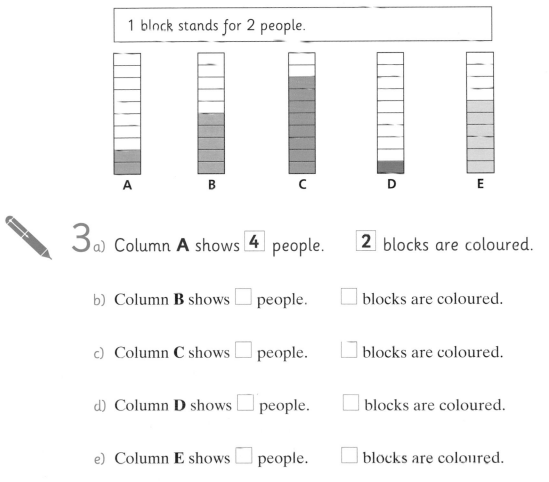

1 block stands for 2 people.

A   B   C   D   E

3 a) Column **A** shows 4 people.   2 blocks are coloured.

b) Column **B** shows ☐ people.   ☐ blocks are coloured.

c) Column **C** shows ☐ people.   ☐ blocks are coloured.

d) Column **D** shows ☐ people.   ☐ blocks are coloured.

e) Column **E** shows ☐ people.   ☐ blocks are coloured.

4   Try Worksheet 1 *Colour the blocks (1)*.

5   Here are the results of a survey about TV soaps.

TV programme	Tally marks
EastEnders	⊞⊞ ⊞⊞ ⊞⊞ I
Coronation Street	⊞⊞ ⊞⊞
Neighbours	⊞⊞ ⊞⊞ ⊞⊞ ⊞⊞
Emmerdale	⊞⊞ III
Home and Away	⊞⊞ ⊞⊞ ⊞⊞ III
Brookside	⊞⊞ ⊞⊞ II

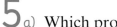

You need

• squared paper, or Worksheet 2.

5 a) Which programme has the biggest number of tally marks?

b) Copy the grid below on to squared paper and complete the block graph, or use Worksheet 2.

TV soaps

```
20 ┌──┬──┬──┬──┬──┬──┐
18 ├──┼──┼──┼──┼──┼──┤
16 ├──┼──┼──┼──┼──┼──┤
14 ├──┼──┼──┼──┼──┼──┤
12 ├──┼──┼──┼──┼──┼──┤
10 ├──┼──┼──┼──┼──┼──┤
 8 ├──┼──┼──┼──┼──┼──┤
 6 ├──┼──┼──┼──┼──┼──┤
 4 ├──┼──┼──┼──┼──┼──┤
 2 ├──┼──┼──┼──┼──┼──┤
 0 └──┴──┴──┴──┴──┴──┘
```

6 Here is a block graph from the results of a survey done with a **different** group of people.

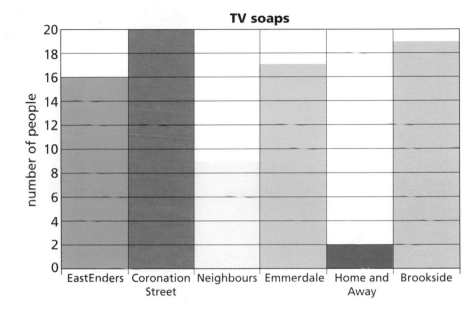

6 a) ☐ people watched Home and Away.

b) ☐ people watched Coronation Street.

c) ☐ people watched EastEnders.

d) How many people do you think watched Neighbours?

e) How many people do you think watched Brookside?

f) How many people do you think watched Emmerdale?

g) Look at the two block graphs in **Questions 5** and **6**.

How are they different?

Why do you think they might be different?

7 How many people do these columns show?

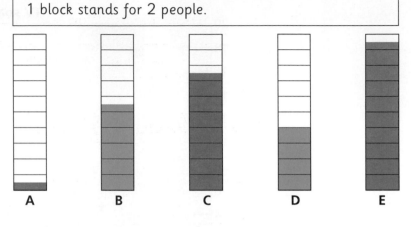

1 block stands for 2 people.

A   B   C   D   E

7 a) Column **A** shows 1 person. **Half** a block is coloured.

b) Column **B** shows ☐ people. _____ blocks are coloured.

c) Column **C** shows ☐ people. _____ blocks are coloured.

d) Column **D** shows ☐ people. _____ blocks are coloured.

e) Column **E** shows ☐ people. _____ blocks are coloured.

8 Try Worksheet 3 *Colour the blocks (2).*

9 You need

• squared paper, or Worksheet 4.

Show **Question 6** as a pictogram.
Draw it on squared paper, or use Worksheet 4.

10 Do your own TV survey in your tutor group.

Make a tally table, then draw your block graph.

You could use Worksheet 5 to help you.

1 block stands for 2 people.

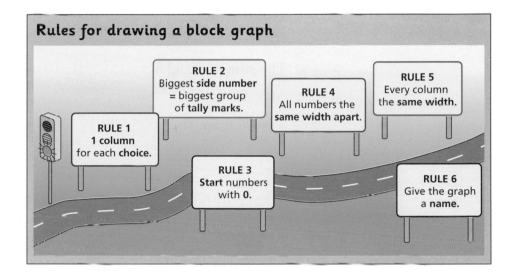

Rules for drawing a block graph

RULE 1
1 column
for each **choice**.

RULE 2
Biggest **side number**
= biggest group
of **tally marks**.

RULE 3
Start numbers
with **0**.

RULE 4
All numbers the
same width apart.

RULE 5
Every column
the **same width**.

RULE 6
Give the graph
a **name**.

Now try Unit 7 Test.

Review 7

1 a) 9 **less than** 10 = ☐   b) 8 **less than** 10 = ☐

c) 10 **less than** 10 = ☐   d) The **sum** of 9 and 9 = ☐

e) The **sum** of 8 and 8 = ☐   f) 10 **more than** 10 is ☐

g) 6 **more than** 0 is ☐

2 What part is coloured, **half** or **quarter**?

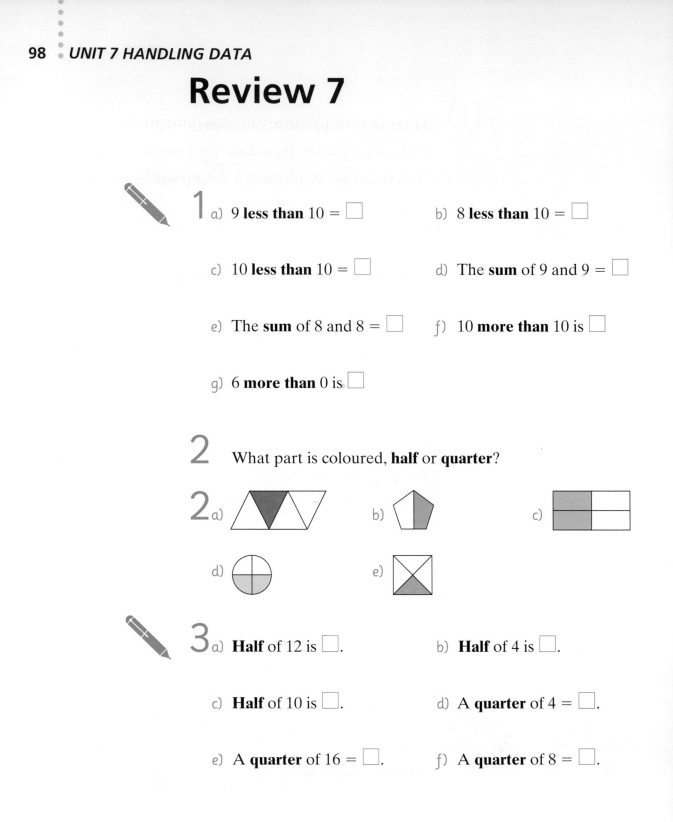

3 a) **Half** of 12 is ☐.   b) **Half** of 4 is ☐.

c) **Half** of 10 is ☐.   d) A **quarter** of 4 = ☐.

e) A **quarter** of 16 = ☐.   f) A **quarter** of 8 = ☐.

4 What are these numbers to the **nearest 10**?

a) 51 is about ☐.　　　　b) 55 is roughly ☐.

c) 78 is about ☐.　　　　d) 62 is about ☐.

e) 96 is roughly ☐.　　　f) 72 is about ☐.

5 a) 100 + 40 + 8 = ☐　　b) 100 + 40 = ☐

c) 100 + 8 = ☐

6 There are 10 eggs in a box.
How many eggs are there in 7 boxes?

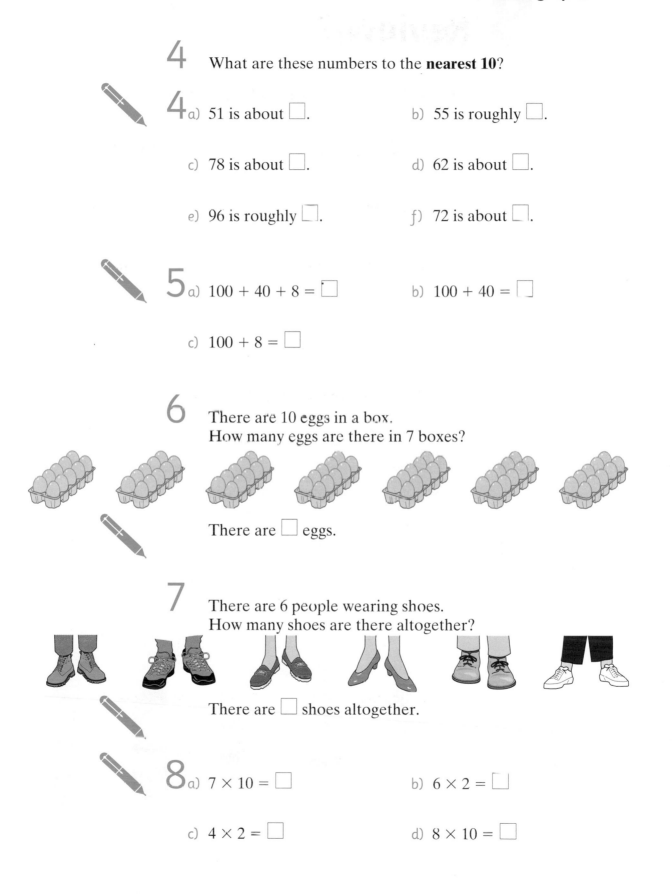

There are ☐ eggs.

7 There are 6 people wearing shoes.
How many shoes are there altogether?

There are ☐ shoes altogether.

8 a) 7 × 10 = ☐　　　　b) 6 × 2 = ☐

c) 4 × 2 = ☐　　　　　d) 8 × 10 = ☐

8 :Number

Subtraction facts to 20

Unit 8 words

take away	subtract	minus
count back	more	less than
most	fewer	makes
half	quarter	symmetry

Remember

Examples are shown in red.

 means copy and complete.

 You need

- a set of Unit 8 vocabulary Snap cards.

 Play a game of Snap to help you learn the words.

Try the **word test** to get some points.

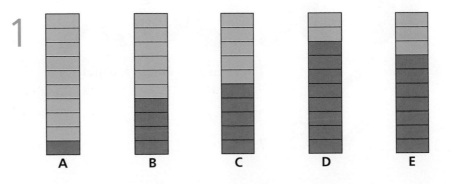

1   A   B   C   D   E

1 a) If you take away the red blocks, how many blocks will be left in each tower?

     9 blocks will be left in tower **A**.

     ☐ blocks will be left in tower **B**.

     ☐ blocks will be left in tower **C**.

     ☐ blocks will be left in tower **D**.

     ☐ blocks will be left in tower **E**.

 b) If you take away the blue blocks, how many blocks will be left in each tower?

     1 block will be left in tower **A**.

     ☐ blocks will be left in tower **B**.

     ☐ blocks will be left in tower **C**.

     ☐ blocks will be left in tower **D**.

     ☐ blocks will be left in tower **E**.

c) Look at each pair of towers.

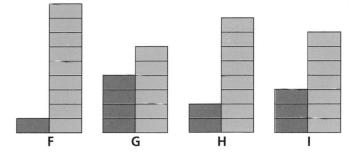

F   G   H   I

There are **8** **more** blue blocks than red blocks in **F**.

There are ☐ **more** blue blocks than red blocks in **G**.

There are ☐ **fewer** red blocks than blue blocks in **H**.

There are ☐ **fewer** red blocks than blue blocks in **I**.

2   Some friends go to the dentist.

They had these numbers of fillings.

Person	Number of fillings
Parminder	0
Bob	4
Tracy	2
Eddie	6
Anne	9

2 a) _____ had **most** fillings.

b) _____ had **fewest** fillings.

c) Anne had ☐ **more** fillings than Eddie.

Eddie had ☐ **more** fillings than Bob.

Bob had ☐ **more** fillings than Parminder.

d) Eddie had ☐ **fewer** fillings than Anne.
Bob had ☐ **fewer** fillings than Eddie.
Parminder had ☐ **fewer** fillings than Bob.

e) If they had each had 2 fillings **fewer**:
Anne would have had ☐ fillings.
Tracy would have had ☐ fillings.
Eddie would have had ☐ fillings.
Bob would have had ☐ fillings.

f) Could Parminder have **2 fewer** fillings?

 3 Try Worksheet 1 *Toothless*.

 4 Copy the number lines and draw in the arrows, or use Worksheet 2.

 4 a) 0 1 2 3 4 5 6 7 8 9 10 11 12 13 14 15 16 17 18 19 20
Count **back 1** from 10 = ☐

b) 0 1 2 3 4 5 6 7 8 9 10 11 12 13 14 15 16 17 18 19 20
Count **back 1** from 20 = ☐

c) 0 1 2 3 4 5 6 7 8 9 10 11 12 13 14 15 16 17 18 19 20
Count **back 3** from 9 = ☐

d) 0 1 2 3 4 5 6 7 8 9 10 11 12 13 14 15 16 17 18 19 20
Count **back 3** from 19 = ☐

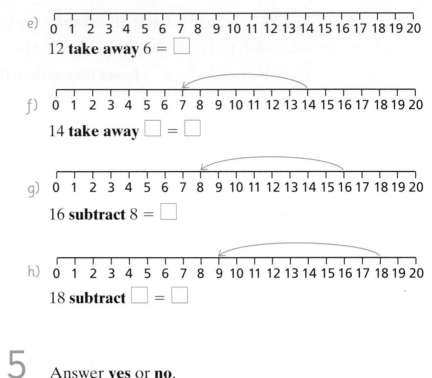

e) 0 1 2 3 4 5 6 7 8 9 10 11 12 13 14 15 16 17 18 19 20
12 **take away** 6 = ☐

f) 0 1 2 3 4 5 6 7 8 9 10 11 12 13 14 15 16 17 18 19 20
14 **take away** ☐ = ☐

g) 0 1 2 3 4 5 6 7 8 9 10 11 12 13 14 15 16 17 18 19 20
16 **subtract** 8 = ☐

h) 0 1 2 3 4 5 6 7 8 9 10 11 12 13 14 15 16 17 18 19 20
18 **subtract** ☐ = ☐

5 Answer **yes** or **no**.

5 a) Can you eat 20 sweets if you have 15 sweets?

a) no

b) Can you eat 12 sweets if you have 5 sweets?

c) Can you eat 9 swcets if you have 18 sweets?

d) Can you eat 10 sweets if you have 20 sweets?

e) Can you take 19 sweets from 9 sweets?

f) Can you take 6 sweets from 12 sweets?

6 Copy these pairs of numbers. Circle the **larger** number.

6 a) 9   14   b) 8   13   c) 19   12

d) 14   17   e) 8   16   f) 12   6

7 In **Question 6**, take the **smaller** numbers from the **larger** numbers.

7 a) 14 − 9 = **5**

8 A shopkeeper gives change in (1p) coins.
She **counts on** to give change.

First Dave spends 2p and pays with a (5p) coin.

2p and 1p make 3p and 1p makes 4p and 1p makes 5p

That gives me 3p change

What **change** would Dave get if he:

8 a) spends 2p and pays with a (5p) coin?

a) 2p + **3p** change makes 5p.

5p − 2p = **3p**

b) spends 3p and pays with a (5p) coin?

3p + ☐ change makes 5p.

5p − 3p = ☐

c) spends 4p and pays with a (10p) coin?

4p + ☐ change makes 10p.

10p − 4p = ☐

d) spends 6p and pays with a (10p) coin?

6p + ☐ change makes 10p.

10p − 6p = ☐

e) spends 5p and pays with a (10p) coin?

5p + ☐ change makes 10p.

10p − 5p = ☐

f) spends 8p and pays with a (20p) coin?

8p + ☐ change makes 20p.

20p − 8p = ☐

g) spends 9p and pays with a (20p) coin?

9p + ☐ change makes 20p.

20p − 9p = ☐

h) spends 12p and pays 15p?

12p + ☐ change makes 15p.

15p − 12p = ☐

9 Tom and Sam save (1p) coins in a bottle.

Tom's bottle   Sam's bottle

9 a) Tom has ☐ (1p) coins.   b) Sam has ☐ (1p) coins.

c) _____ has **most** (1p) coins.

He has ☐ **more** coins.

d) _____ has **fewest** (1p) coins.

He has ☐ **fewer** coins.

e) The **difference** between the number of coins in Tom's bottle
and the number of coins in Sam's bottle is ☐.

Remember

To find the **difference**, we **take away**.
We take the **smaller** number from the **larger** number.

10 Find the **difference** between:

10 a) 2p and 3p   b) 12p and 13p   c) 10p and 1p

d) 20p and 1p   e) 18p and 10p   f) 18p and 9p

g) 6p and 12p   h) 7p and 14p   i) 16p and 8p.

j) How many pairs of numbers can you find
between 0 and 20 with a **difference of 10**?

11 Try Worksheet 3 *Find the difference.*

12 You need

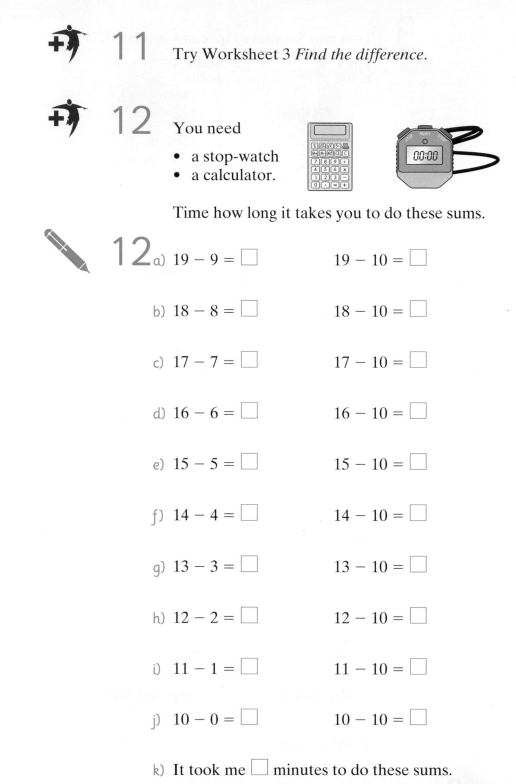

- a stop-watch
- a calculator.

Time how long it takes you to do these sums.

12

a) $19 - 9 = \square$ $19 - 10 = \square$

b) $18 - 8 = \square$ $18 - 10 = \square$

c) $17 - 7 = \square$ $17 - 10 = \square$

d) $16 - 6 = \square$ $16 - 10 = \square$

e) $15 - 5 = \square$ $15 - 10 = \square$

f) $14 - 4 = \square$ $14 - 10 = \square$

g) $13 - 3 = \square$ $13 - 10 = \square$

h) $12 - 2 = \square$ $12 - 10 = \square$

i) $11 - 1 = \square$ $11 - 10 = \square$

j) $10 - 0 = \square$ $10 - 10 = \square$

k) It took me $\square$ minutes to do these sums.

l) Check your answers on the calculator.

 **13** Can you think of a quick way to help you **take 9** from numbers up to 20?

14 Mike and Tina go shopping. There is a sale on.

Before the sale:

14 a) the jacket costs £☐ more than the shirt

b) the trainers cost £☐ more than the scarf

c) the shorts cost £☐ more than the scarf

d) the trainers cost £☐ more than the jumper.

The sale price of:

e) the scarf is £☐

f) the trainers is £☐

g) the shirt is £☐

h) the jumper is £☐

i) the jacket is £☐

j) the shorts is £☐

15 What is **half** of these prices?

Remember

To get a **half** we divide into **2 equal** groups.

15 a) Half of £6 is £ 3 .

b) Half of £12 is £☐.

c) Half of £14 is £☐.

d) Half of £16 is £☐.

e) Half of £18 is £☐.

f) Half of £20 is £☐.

g) If all the clothes in **Question 14** had been **half price** in the sale, how much would each have cost?

16

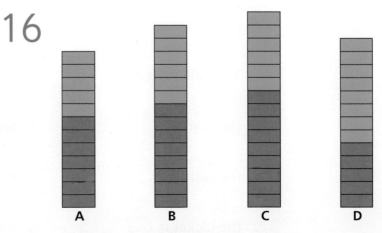

Look at the towers to help you.

16 a) You have **seven** CDs.
You need to buy _____ more to make **twelve**.

b) You have **eight** CDs.
You need to buy _____ more to make **fourteen**.

c) You have **nine** CDs.
You need to buy _____ more to make **fifteen**.

d) You have **five** CDs.
You need to buy _____ more to make **thirteen**.

17 Look at the towers in **Question 16** to help you.

17 a) $7 + 5 = \Box$
$5 + 7 = \Box$
$12 - 5 = \Box$
$12 - 7 = \Box$

b) $8 + 6 = \Box$
$6 + 8 = \Box$
$14 - 8 = \Box$
$14 - 6 = \Box$

c) $9 + 6 = \Box$
$6 + 9 = \Box$
$15 - 9 = \Box$
$15 - 6 = \Box$

d) $8 + 5 = \Box$
$5 + 8 = \Box$
$13 - 8 = \Box$
$13 - 5 = \Box$

 18 Try Worksheet 4 *Check it out.*

 **19** Make up some cartoon problems.
Use Worksheet 5 *Problem page*.

 **20** Try Worksheet Puzzle *Take-away moves*.

21 You need

- Unit 8 Race against time cards (numbers and words)
- Unit 1 Race against time cards (numbers and words)
- your 'My maths record' sheet.

Race against time

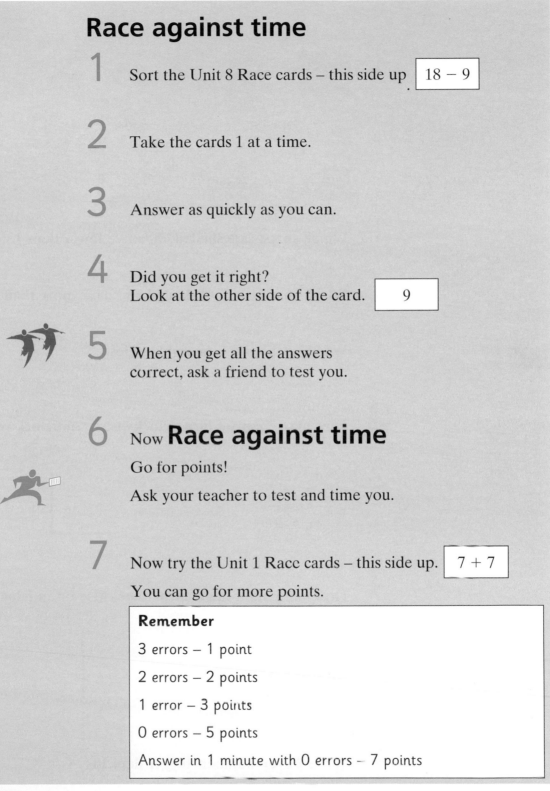

1 Sort the Unit 8 Race cards – this side up. | 18 − 9 |

2 Take the cards 1 at a time.

3 Answer as quickly as you can.

4 Did you get it right?
Look at the other side of the card. | 9 |

5 When you get all the answers correct, ask a friend to test you.

6 Now **Race against time**

Go for points!

Ask your teacher to test and time you.

7 Now try the Unit 1 Race cards – this side up. | 7 + 7 |

You can go for more points.

> **Remember**
>
> 3 errors – 1 point
>
> 2 errors – 2 points
>
> 1 error – 3 points
>
> 0 errors – 5 points
>
> Answer in 1 minute with 0 errors – 7 points

Now try Unit 8 Test.

Review 8

1

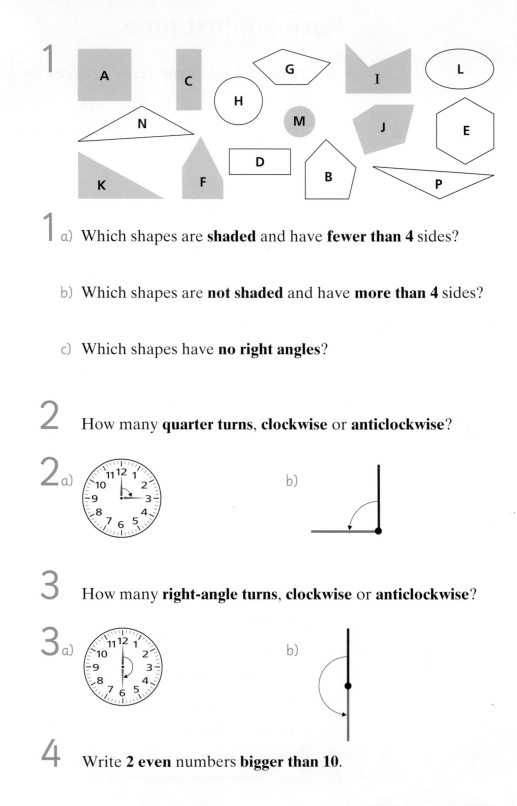

1 a) Which shapes are **shaded** and have **fewer than 4** sides?

b) Which shapes are **not shaded** and have **more than 4** sides?

c) Which shapes have **no right angles**?

2 How many **quarter turns**, **clockwise** or **anticlockwise**?

2 a)   b)

3 How many **right-angle turns**, **clockwise** or **anticlockwise**?

3 a)   b)

4 Write **2 even** numbers **bigger than 10**.

5 Write down:

5 a) one thing that measures **more than** a metre

   b) one thing that measures **less than** a metre

   c) one thing that measures **about** a metre.

9 Measures

Estimating and measuring length

Unit 9 words

metre	centimetre	measure
length	longest	shortest
tall	roughly	count back
sides	block graph	worth

Remember

Examples are shown in red.

means copy and complete.

You need
- a set of Unit 9 vocabulary Snap cards.

Play a game of Snap to help you learn the words.

Try the **word test** to get some points.

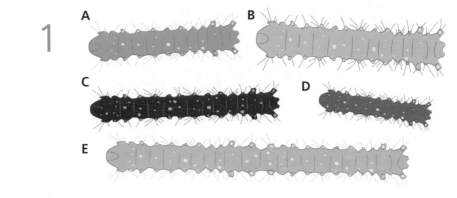

1

1 a) Caterpillar **A** is ☐ cm long.

b) Caterpillar **B** is ☐ cm long.

c) Caterpillar **C** is ☐ cm long.

d) Caterpillar **D** is ☐ cm long.

e) Caterpillar **E** is ☐ cm long.

f) Caterpillar _____ is the **longest**.

g) Caterpillar _____ is the **shortest**.

h) Caterpillars _____ and _____ are the **same length**.

2 a) Draw a line any length in whole centimetres up to 20 cm.

Remember

Use your ruler like this.

start here

0 1 2 3 4 5 6 7 8 9

b) Your partner guesses the length and writes the guess on your line.

c) Your partner then measures the line and writes the real length on it.

guess 6 cm   length 7 cm

d) Take turns to draw 6 lines each. Guess and measure the length of your partner's lines.

3

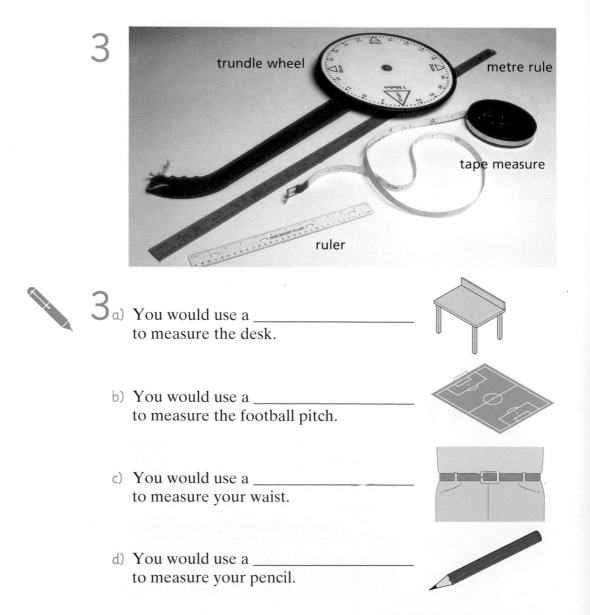

trundle wheel

metre rule

tape measure

ruler

3 a) You would use a _____ to measure the desk.

b) You would use a _____ to measure the football pitch.

c) You would use a _____ to measure your waist.

d) You would use a _____ to measure your pencil.

4 To help you guess centimetre lengths.

4 a) Find one thing that is roughly 1 cm long.

_____ is roughly 1 cm long.

b) Find one thing that is roughly 10 cm long.

_____ is roughly 10 cm long.

c) Find one thing that is roughly 30 cm long.

_____ is roughly 30 cm long.

Remember

Remember these things as your own **centimetre reminders**.

My own centimetre reminders are _____,

_____ and _____ .

5 What are these numbers **roughly** – to the **nearest ten**?

5 a) 12 is between 10 and 20.

The middle number is 15.

12 is nearer to 10.

We say 12 is **roughly** 10.

b) 14 is between ☐ and ☐.

   The middle number is ☐.

   14 is nearer to ☐.

   14 is **roughly** ☐.

c) 18 is between ☐ and ☐.

   The middle number is ☐.

   18 is **roughly** ☐.

d) 13 is between ☐ and ☐.

   13 is **about** ☐.

e) 16 is between ☐ and ☐.

   16 is **about** ☐.

f) 11 is **about** ☐.

g) 17 is **roughly** ☐.

h) 15 is in the middle.

   It is **roughly** ☐.

6 Tutor group 8C is growing plants from seed in Science.
   Jenny wants to know how tall Bob's plant has grown.
   She does not want to know how tall it is **exactly**.
   She wants to know **roughly**.

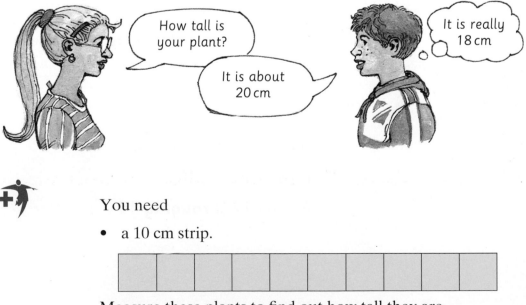

How tall is your plant?

It is about 20 cm

It is really 18 cm

You need

• a 10 cm strip.

Measure these plants to find out how tall they are.
Use the 10 cm strip to help you.

6 a) Plant **A** is 12 cm tall.

12 cm is between 10 cm and 20 cm.

The middle number is 15 cm.

Plant **A** is **roughly** 10 cm.

b) Plant **B** is ☐ cm tall.

Plant **B** is **roughly** ☐ cm.

c) Plant **C** is ☐ cm tall.

Plant **C** is **roughly** ☐ cm.

d) Plant **D** is ☐ cm tall.

Plant **D** is **about** ☐ cm.

e) Plant **E** is ☐ cm tall.

Plant **E** is **about** ☐ cm.

f) The **tallest** plant is plant _____.

g) The **shortest** plant is plant _____.

h) The **difference** in their heights is ☐ cm.

7 Try Worksheet 1 *A rough measure*.

8 What are these measurements **roughly** – to the **nearest ten**?

8 a) 11 cm is **roughly** ☐ cm.

b) 21 cm is **roughly** ☐ cm.

c) 32 cm is **roughly** ☐ cm.

d) 53 cm is **roughly** ☐ cm.

e) 74 cm is **roughly** ☐ cm.

f) 75 cm is **roughly** ☐ cm.

g) 16 cm is **about** ☐ cm.

h) 26 cm is **about** ☐ cm.

i) 37 cm is **about** ☐ cm.

j) 58 cm is **about** ☐ cm.

k) 79 cm is **about** ☐ cm.

l) 99 cm is **about** ☐ cm.

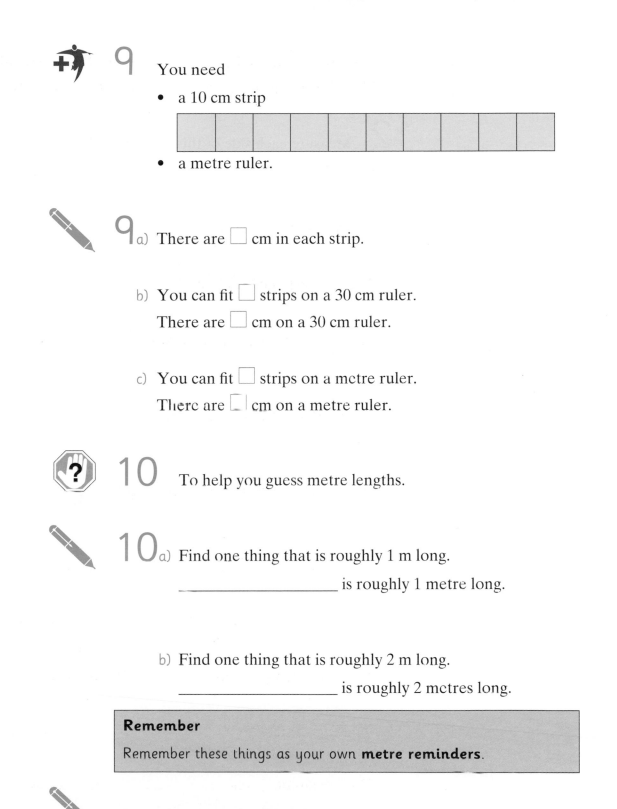

9   You need
   • a 10 cm strip

   • a metre ruler.

9 a) There are ☐ cm in each strip.

   b) You can fit ☐ strips on a 30 cm ruler.
      There are ☐ cm on a 30 cm ruler.

   c) You can fit ☐ strips on a metre ruler.
      There are ☐ cm on a metre ruler.

10   To help you guess metre lengths.

10 a) Find one thing that is roughly 1 m long.
      _____ is roughly 1 metre long.

   b) Find one thing that is roughly 2 m long.
      _____ is roughly 2 metres long.

> **Remember**
> Remember these things as your own **metre reminders**.

My own metre reminders are _____ and
_____ .

11 Try Worksheet 2 *Check it out*.

12 Measure each side of this shape.

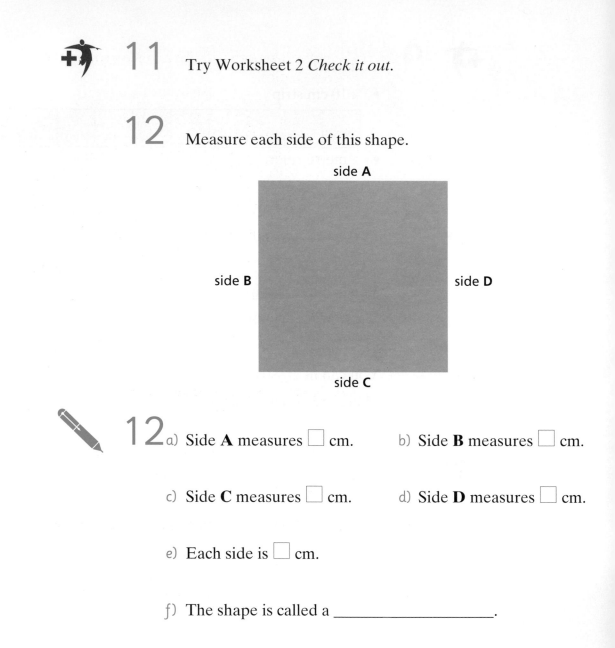

side **A**

side **B**

side **D**

side **C**

12

a) Side **A** measures ☐ cm. b) Side **B** measures ☐ cm.

c) Side **C** measures ☐ cm. d) Side **D** measures ☐ cm.

e) Each side is ☐ cm.

f) The shape is called a _____.

13 Cal helps his mother build a fence around the vegetable plot.

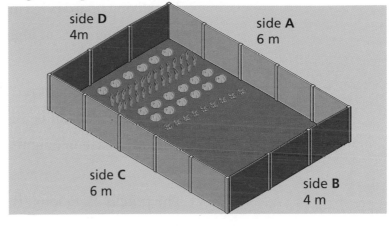

side **D**
4m

side **A**
6 m

side **C**
6 m

side **B**
4 m

13 a) Side A is ☐ m long.

☐ m long.

☐ m **longer than** side **B**.

the plot is a _____ .

holiday to this apartment block.

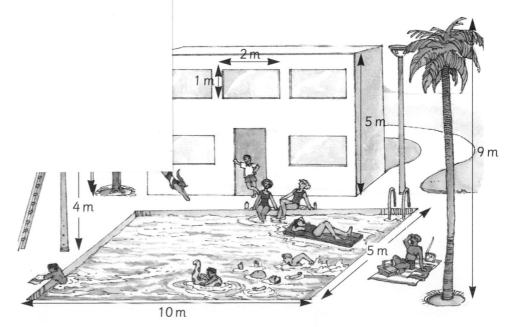

2 m

1 m

5 m

9 m

4 m

5 m

10 m

14 a) The building is ☐ m high.

The diving board is ☐ m high.

The building is ☐ m **higher than** the diving board.

b) The tree on the right is ☐ m high.

The tree on the left is ☐ m high.

The **difference** in heights is ☐ m.

c) The pool is ☐ m long. It is ☐ m wide.

The pool is ☐ m **longer** than it is **wide**.

d) The length of **2** windows would be ☐ m.

The length of **10** windows would be ☐ m.

e) I guess that the door is about ☐ m high.

f) I guess that the small boy by the door is roughly ☐ m tall.

g) I guess that the light is about ☐ m high.

h) The length of all 4 sides of one window is ☐ m.

15 The graph shows the size of the **hand spans** of a group of people.

17 cm

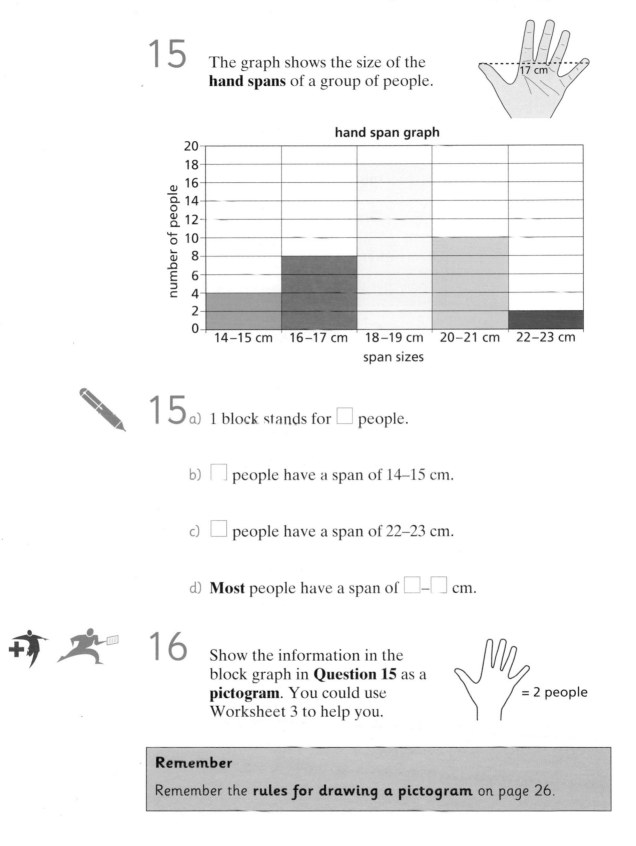

hand span graph

15 a) 1 block stands for ☐ people.

b) ☐ people have a span of 14–15 cm.

c) ☐ people have a span of 22–23 cm.

d) **Most** people have a span of ☐–☐ cm.

16 Show the information in the block graph in **Question 15** as a **pictogram**. You could use Worksheet 3 to help you.

= 2 people

Remember

Remember the **rules for drawing a pictogram** on page 26.

17 You need

- a 30 cm ruler
- some squared paper.

Do your own survey of the hand spans of people in your tutor group.

Draw a **block graph** to show your results.

> **Remember**
>
> Remember the **rules for drawing a block graph** on page 22.

Copy the tally table below to help you, or use Worksheet 4.

Hand span	Tally marks
14–15 cm	
16–17 cm	
18–19 cm	
20–21 cm	
22–23 cm	

Class guess

18 Play a game of 'Class guess' each day for a week.

Rules for 'Class guess'

1 Someone holds up or points to an object.
2 Everyone writes down their guess of its length.
3 Two people collect up the written guesses.
4 The person who chose the object measures its length.
5 Your teacher will tell you how near the guess has to be to be a 'good' guess.
6 Keep score of the 'good' guesses:

Monday – **3** 'good' guesses.

7 At the end of the week, draw a **block graph**.

Put 5 columns along the bottom, for Monday to Friday.

Are there more 'good' guesses as the week goes on?

Now try Unit 9 Test.

Review 9

1 Here is a block graph showing the number of pairs of shoes sold in a shop.

Copy and complete the table.

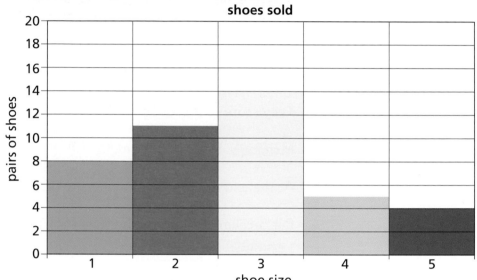

Shoe size	Number sold
1	
2	
3	
4	
5	

1 a) The most popular shoe size was size ☐.

b) The least popular shoe size was size ☐.

2 a) $3 \times 2 = \square$ b) $8 \times 10 = \square$ c) $3 \times 10 = \square$

d) $6 \times 2 = \square$ e) $4 \times 10 = \square$ f) $8 \times 2 = \square$

g) $6 \times 10 = \square$ h) $5 \times 2 = \square$ i) $7 \times 10 = \square$

j) $2 \times 2 = \square$ k) $9 \times 10 = \square$ l) $4 \times 2 = \square$

m) $2 \times 10 = \square$ n) $9 \times 2 = \square$ o) $5 \times 10 = \square$

p) $7 \times 2 = \square$

3 Write down the number **before** and the number **after** these numbers.

3 a) $\square$ **69** $\square$ b) $\square$ **150** $\square$

c) $\square$ **200** $\square$ d) $\square$ **101** $\square$

4 What are these numbers to the **nearest ten**?

4 a) 32 b) 38 c) 35 d) 81 e) 89 f) 85

5 a) Half of six apples = _____ apples.

b) Quarter of twelve apples = _____ apples.

c) Quarter of four apples = _____ apple.

10 Number

Place value to 500

Unit 10 words

digit	hundred	tens
units	nearer to	value
altogether	single	greater than
less than	hexagon	pentagon

Remember

Examples are shown in red.

 means copy and complete.

 You need
* a set of Unit 10 vocabulary Snap cards.

 Play a game of Snap to help you learn the words.

Try the **word test** to get some points.

1 Write these numbers in **words**.
What is the value of the circled digit – **hundreds**, **tens** or **units**?

1 a) ①4   a) **fourteen – 1 ten**

b) ①0 4   c) 9①   d) 2 0①

e) (2)9   f) 2(2)   g) (2)8 2   h) 1(2)0

2   Which tens are these numbers between?

2 a) 16 is between **ten** and **twenty**.

b) 26 is between _____ and _____.

c) 62 is between _____ and _____.

d) 74 is between _____ and _____.

e) 88 is between _____ and _____.

f) 97 is between _____ and _____.

3   What are these numbers **roughly** – to the **nearest ten**?

3 a) 32 is roughly **thirty**.

b) 72 is roughly _____.

c) 85 is roughly _____.

d) 67 is roughly _____.

e) 99 is about _____.

f) 29 is about _____.

g) 95 is about _____.

h) 55 is about _____.

4 There are **10 biscuits** in a **packet**.

There are **10 packets** of biscuits in a **box**.

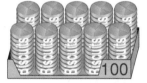

4 a) There are ☐ biscuits in one packet.

b) There are ☐ biscuits altogether in one box.

c) There are ☐ units in one ten.

d) There are ☐ tens in one hundred.

How many biscuits are there altogether in each drawing?

e)

e) There are **2** hundreds **9** tens **8** units.
There are **298** biscuits altogether.

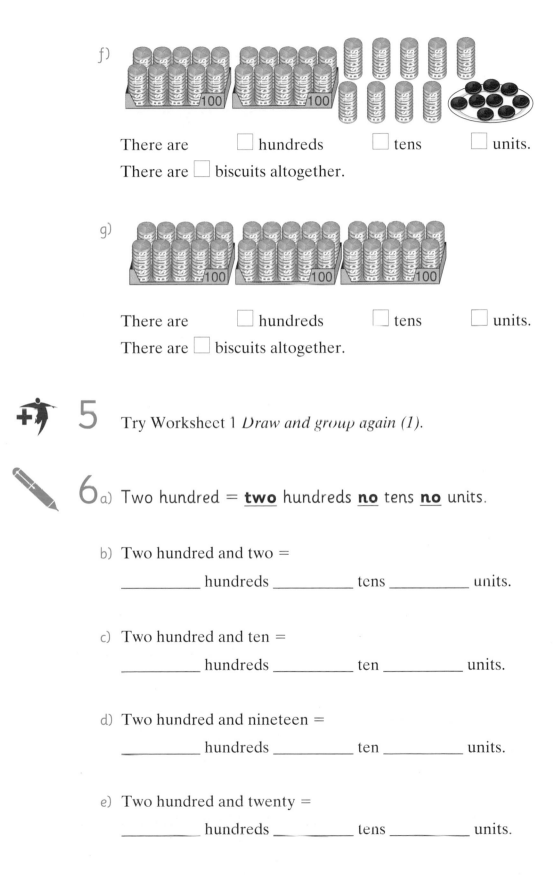

f)

There are ☐ hundreds ☐ tens ☐ units.

There are ☐ biscuits altogether.

g)

There are ☐ hundreds ☐ tens ☐ units.

There are ☐ biscuits altogether.

5 Try Worksheet 1 *Draw and group again (1)*.

6 a) Two hundred = **two** hundreds **no** tens **no** units.

b) Two hundred and two =

_____ hundreds _____ tens _____ units.

c) Two hundred and ten =

_____ hundreds _____ ten _____ units.

d) Two hundred and nineteen =

_____ hundreds _____ ten _____ units.

e) Two hundred and twenty =

_____ hundreds _____ tens _____ units.

7 How many biscuits are there **altogether** in each drawing?

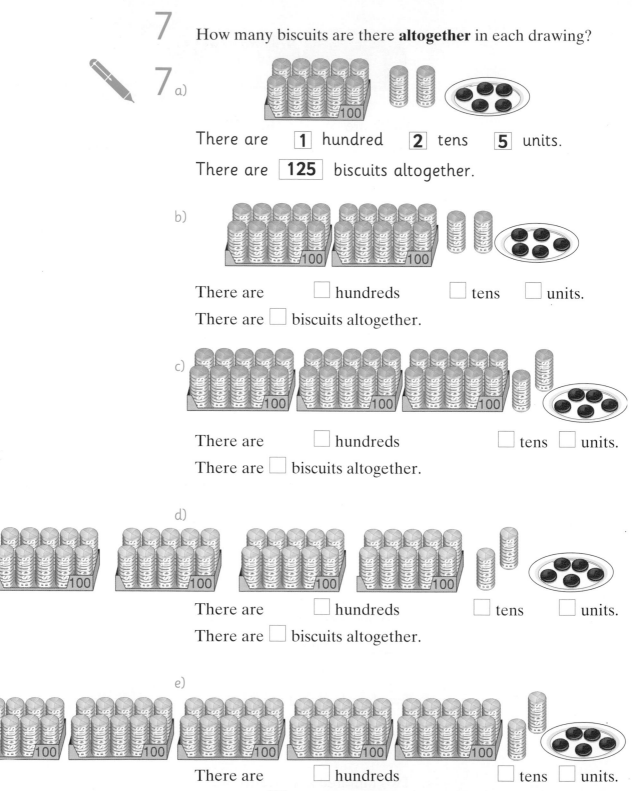

7 a)

There are   **1** hundred   **2** tens   **5** units.

There are   **125**   biscuits altogether.

b)

There are ☐ hundreds   ☐ tens   ☐ units.

There are ☐ biscuits altogether.

c)

There are ☐ hundreds   ☐ tens ☐ units.

There are ☐ biscuits altogether.

d)

There are ☐ hundreds   ☐ tens   ☐ units.

There are ☐ biscuits altogether.

e)

There are ☐ hundreds   ☐ tens ☐ units.

There are ☐ biscuits altogether.

8 Try Worksheet 2 *Draw and group again (2)*.

9 Write these numbers in **words**.

What is the value of the circled digit – **hundreds**, **tens** or **units**?

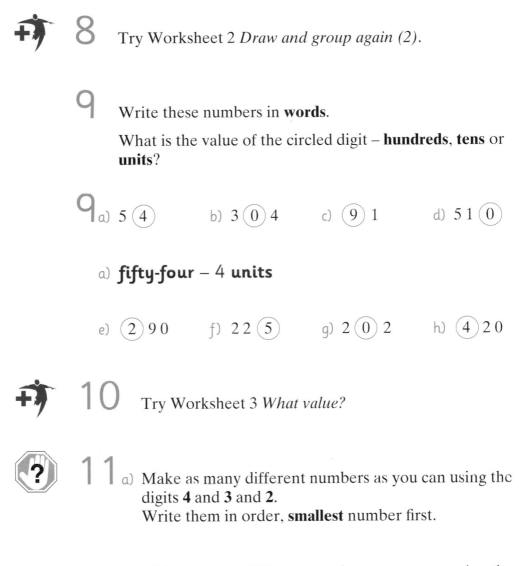

a) 5 (4)　　　b) 3 (0) 4　　　c) (9) 1　　　d) 5 1 (0)

a) **fifty-four** – 4 **units**

e) (2) 9 0　　　f) 2 2 (5)　　　g) 2 (0) 2　　　h) (4) 2 0

10 Try Worksheet 3 *What value?*

11 a) Make as many different numbers as you can using the digits **4** and **3** and **2**.
Write them in order, **smallest** number first.

b) Make as many different numbers as you can using the digits **5** and **3** and **1**.
Write them in order, **smallest** number first.

12 See if your teacher wants you to play a game of 'Cover up'.

You need

• 'Cover up' cards.

Look at these 3 cards.

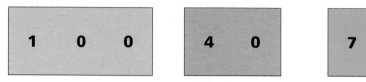

Put together, they can make a **3-digit** number.

12 a) How many digits are covered in the ⌈100⌉ card?

☐ digits are covered.

b) How many digits are covered in the ⌈40⌉ card?

☐ digit is covered.

13 Look at these 2 cards.

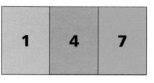

Put together, they can make a **3-digit** number.

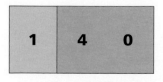

13 a) How many digits are covered in the ☐100 card?

☐ digits are covered.

b) How many digits are covered in the ☐40 card?

_____ digits are covered.

 c) Why are **no** digits covered in the ☐40 card this time?

14 Look at these 2 cards.

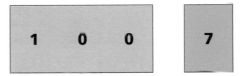

Put together, they can make a **3-digit** number.

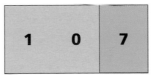

 14 a) How many digits are covered in the ☐100 card?

☐ digit is covered.

 b) Why is only **one** digit covered in the ☐100 card this time?

15

Draw these cards so that they make a **3-digit** number.

Write the number.

To help you, look at the cards in **Questions 12**, **13** and **14**.

15 a)

| 2 0 0 | 4 0 | 5 |

b)

| 2 0 0 | 4 0 |

c)

| 2 0 0 | 5 |

d)

| 3 0 0 | 5 0 | 7 |

e)

| 4 0 0 | 2 0 |

f)

| 5 0 0 | 1 |

16

How many biscuits?

16 a)

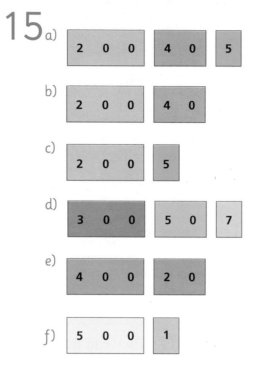

a) There are [**123**] biscuits

b)

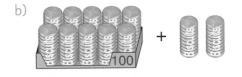

There are ☐ biscuits.

c)

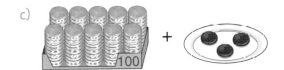

There are ☐ biscuits.

d)

There are ☐ biscuits.

e)

There arc ☐ biscuits.

f)

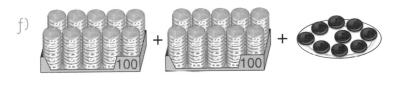

There are ☐ biscuits.

17 Find the **total**.

> **Remember**
>
> **Total** means **add**.

17 a) $100 + 20 + 3 = \boxed{123}$ b) $100 + 20 = \boxed{120}$ c) $100 + 3 = \boxed{103}$

d) $100 + 30 + 9 = \square$ e) $100 + 30 = \square$ f) $100 + 9 = \square$

g) $200 + 50 + 8 = \square$ h) $200 + 50 = \square$ i) $200 + 8 = \square$

j) $800 + 10 + 4 = \square$ k) $800 + 10 = \square$ l) $800 + 4 = \square$

18 Put these digits in the correct place.

18 a) 123

> **Remember**
>
> **H** = hundreds   **T** = tens   **U** = units

a)

hundreds	tens	units
1	2	3

b) 120

hundreds	tens	units

c) 103

H	T	U

d) 249

H	T	U

e) 240

hundreds	tens	units

f) 209

hundreds	tens	units

19 Find the total.

19 a)
```
  H T U
  1 0 0
+   2 0
──────
  1 2 0
```

b)
```
H T U
  1 0 0
+     3
┌──────┐
└──────┘
```

c)
```
H T U
  1 0 0
+   2 3
┌──────┐
└──────┘
```

d)
```
H T U
  2 0 0
+   4 0
┌──────┐
└──────┘
```

e)
```
H T U
  2 0 0
+     9
┌──────┐
└──────┘
```

f)
```
H T U
  2 0 0
+   4 9
┌──────┐
└──────┘
```

20

Ella wants to know how many football cards Rani has collected.
She does not want to know the **exact** number.
She wants to know **roughly**.

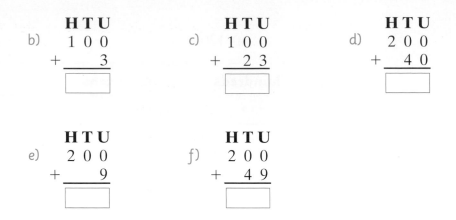

How many football cards have you collected?

I have roughly 130 football cards

I really have 132 football cards

Look at the number line.
Look at the **middle** number.

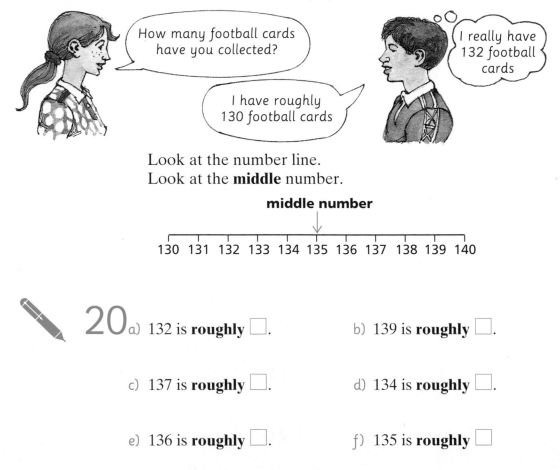

middle number

130 131 132 133 134 135 136 137 138 139 140

20 a) 132 is **roughly** ☐.

b) 139 is **roughly** ☐.

c) 137 is **roughly** ☐.

d) 134 is **roughly** ☐.

e) 136 is **roughly** ☐.

f) 135 is **roughly** ☐

What are these numbers roughly – to the **nearest ten**?

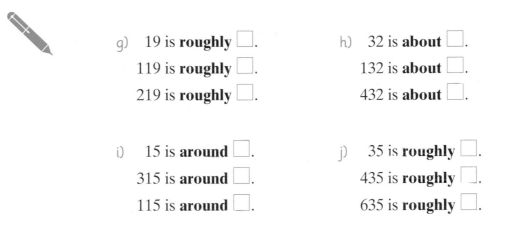

g) 19 is **roughly** ☐.
119 is **roughly** ☐.
219 is **roughly** ☐.

h) 32 is **about** ☐.
132 is **about** ☐.
432 is **about** ☐.

i) 15 is **around** ☐.
315 is **around** ☐.
115 is **around** ☐.

j) 35 is **roughly** ☐.
435 is **roughly** ☐.
635 is **roughly** ☐.

21 Try Worksheet 4 *Round up*.

22 Draw these doors.

Fill in the number **before** the numbered door.
Fill in the number **after** the numbered door.

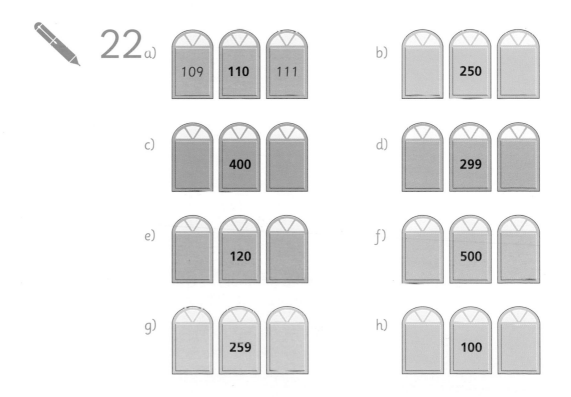

23 You need

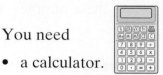

• a calculator.

Your partner puts these numbers into the calculator.
You guess the answer each time.
Your partner shows you the answer on the calculator.
Take turns with each list.

List A	**List B**	**List C**
10 plus 10	100 + 100	190 + 1
110 plus 10	170 + 100	190 + 10
210 plus 10	380 + 100	290 + 10
50 plus 10	230 + 100	390 + 10
150 plus 10	60 + 100	190 + 100
350 plus 10	410 + 100	290 + 100
30 plus 10	301 + 100	390 + 100
130 plus 10	150 + 100	490 + 10
330 plus 10	120 + 100	100 + 1
390 plus 10	400 + 100	1 + 100

24 Write these numbers in order, **smallest** number first.

24 a) 60   260   460   360   160

b) 60   420   230   70   90

c) 325   25   5   125   425

d) 8   149   67   262   349

25 Use the numbers in **Question 24** to play a game of
'Say it, write it'. The rules are on page 34.

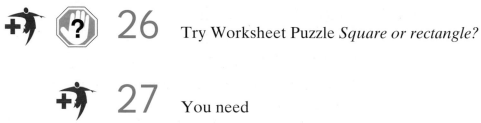

26 Try Worksheet Puzzle *Square or rectangle?*

27 You need
- Unit 10 Race against time cards Set 1 and Set 2
- your 'My maths record' sheet.

Race against time

1   Sort the Set 1 Race cards – this side up.   4 ③ 2

     (Do not mix up Set 1 and Set 2.)

2   Take the cards 1 at a time.

3   Answer as quickly as you can.

4   Did you get it right?

     Look at the other side of the card.   3 tens

5   When you get all the answers correct, ask a friend to test you.

6   Now **Race against time**

     Go for points!

     Ask your teacher to test and time you.

7   Now try Set 2 – this side up.   to nearest ten 358

8   You can go for more points.

Remember
3 errors – 1 point
2 errors – 2 points
1 error – 3 points
0 errors – 5 points
Answer in 1 minute with 0 errors – 7 points

Now try Unit 10 Test.

Review 10

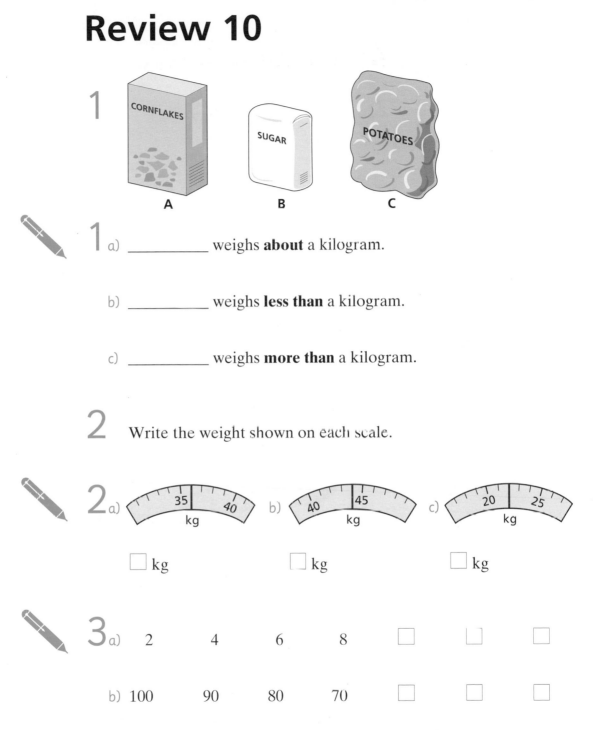

1a) _____ weighs **about** a kilogram.

b) _____ weighs **less than** a kilogram.

c) _____ weighs **more than** a kilogram.

2 Write the weight shown on each scale.

2a) 35 40 kg b) 40 45 kg c) 20 25 kg

☐ kg   ☐ kg   ☐ kg

3a) 2   4   6   8   ☐ ☐ ☐

b) 100   90   80   70   ☐ ☐ ☐

4 What would you have left if you:

a) had 20p and spent 7p?

b) had 14p and spent 8p?

c) had 11p and spent 8p?

d) had 20p and spent 10p?

5 Choose one of these words to complete each sentence.

cube sphere cuboid cylinder

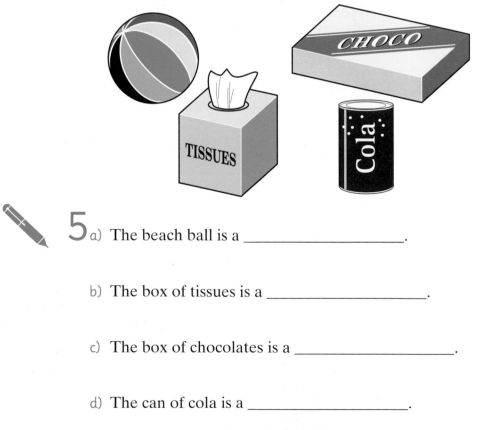

a) The beach ball is a _____.

b) The box of tissues is a _____.

c) The box of chocolates is a _____.

d) The can of cola is a _____.

11 Measures

Estimating and weighing

Unit 11 words

kilogram	left	right
weigh	heaviest	lightest
gram	about	digit
bigger	smaller	tall

Remember

Examples are shown in red.

means copy and complete.

 You need

• a set of Unit 11 vocabulary Snap cards.

 Play a game of Snap to help you learn the words.

 Try the **word test** to get some points.

1

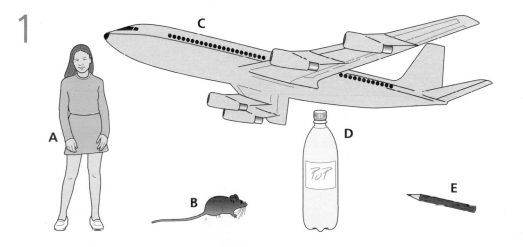

Look at the drawings on the previous page.

> **Remember**
>
> A bag of sugar weighs about 1 kilogram (1 kg).

Choose one of these phrases to complete each sentence.

more than a kilogram   **less than a kilogram**
about a kilogram

1 a) **A** weighs **more than a kilogram**.

b) **B** weighs _____.

c) **C** weighs _____.

d) **D** weighs _____.

e) **E** weighs _____.

f) Find something that weighs **about a kilogram**.

> **Remember**
>
> Remember the object as your own **kilogram reminder**.

My own kilogram reminder is _____.

2 What do these fruits weigh?

a) The bananas weigh ☐ kg. b) The apples weigh ☐ kg.

c) The pears weigh ☐ kg. d) The apples weigh ☐ kg.

e) The plums weigh ☐ kg. f) The bananas weigh ☐ kg.

g) The pears weigh ☐ kg. h) The grapes weigh ☐ kg.

3 Try Worksheet 1 *Weigh in*.

4 All of these things weigh less than a kilogram. Which one weighs about **half** ($\frac{1}{2}$) a kilogram?

The _____ weighs about half a kilogram.

5 Try Worksheet 2 *About half a kilogram.*

6 Find one thing that weighs about half a kilogram.

 _____ weighs roughly half a kilogram.

 7 You need

- a set of books (all the same)
- a set of balance scales
- a half-kilogram weight
- a kilogram weight.

Take turns to guess how many books weigh half a kilogram.

7 a) I guess that ☐ books weigh half a kilogram.

Use the balance scales to check your guess.

b) ☐ books weigh half a kilogram.

c) Guess how many books will weigh a kilogram.

Use the balance scales to check your guess.

d) Find out how many half kilograms there are in one kilogram.

We weigh things that are smaller than a kilogram in **grams** (**g** for short).

$\frac{1}{2}$ kg can also be written as 500 g.

8

We weigh things that are smaller than a kilogram in **grams** (**g** for short).

Look at the foods in the picture.

Choose one of these phrases to complete each sentence.

**more than $\frac{1}{2}$ kg (500 g)   less than $\frac{1}{2}$ kg (500 g)
about $\frac{1}{2}$ kg (500 g)**

8 a) The cornflakes weigh **about $\frac{1}{2}$ kg (500 g)**.

b) The jam weighs _____.

c) The bottle of orange weighs _____.

d) The jelly weighs _____.

e) The beans weigh _____.

f) The curry paste weighs _____.

9 You need

- a calculator.

These are some things that you might eat for lunch.

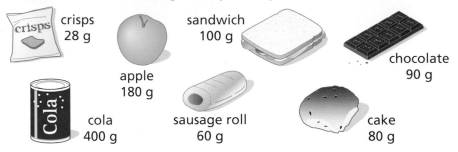

crisps
28 g

apple
180 g

sandwich
100 g

chocolate
90 g

cola
400 g

sausage roll
60 g

cake
80 g

Make up as many different lunches as you can that weigh
about ½ kg (500 g).

Use your calculator to help you.

Remember

½ kg = 500 g

At the greengrocer

Rose

Tony

Mary

10 a) Rose bought ☐ kg of carrots.

b) Tony bought ☐ kg of onions.

c) Mary bought ☐ kg of potatoes.

d) They bought ☐ kg of vegetables **altogether**.

e) _____ has the **heaviest** bag of vegetables.

f) _____ has the **lightest** bag of vegetables.

g) The **difference** in weight between these two bags is ☐ kg.

h) If Mary bought **two times** as many vegetables,

   they would weigh ☐ kg.

i) If Tony bought **half** as many vegetables,

   they would weigh ☐ kg.

11 Tutor group 9D weighed its school bags.
This is what it found.

weights of 9D's bags

11 a) 1 block stands for ☐ bags.

b) ☐ bags weigh 4–7 kg.

c) ☐ bags weigh 8–11 kg.

d) **Most** bags weigh ☐–☐ kg.

12 Show the information in the block graph in **Question 11** as a **pictogram**. You could use Worksheet 3 to help you.

 = 2 bags

> **Remember**
>
> Remember the **rules for drawing a pictogram** on page 26.

13 You need

• a spring balance.

Do your own survey of the **weights of bags** in your tutor group.

Draw a **block graph** to show your results.

> **Remember**
>
> Remember the **rules for drawing a block graph** on page 22.

Decide how many bags each block will stand for.

Copy the tally table below to help you, or use Worksheet 4.

Weight of bag	Tally marks
0–3 kg	
4–7 kg	
8–11 kg	
12–15 kg	
16 kg or more	

Class guess

14 Play a game of 'Class guess' each day for a week.

Rules for 'Class guess'

1 An object is passed around the room.
2 Everyone writes down their guess of its weight.
3 Two people collect up the written guesses.
4 One person weighs the object.
5 Your teacher will tell you how near the guess has to be to be a 'good' guess.
6 Keep score of the 'good' guesses.

   Monday – **6** 'good' guesses.

7 At the end of the week, draw a **block graph**.

   Put 5 columns along the bottom, for Monday to Friday.
   Put the numbers for 'good' guesses up the side.

Are there more 'good' guesses as the week goes on?

Now try Unit 11 Test.

Review 11

1 a) My **30 cm reminder** is _____.

b) My **1 m reminder** is _____.

c) There are ☐ cm in 1 m.

2 What are these measurements **to the nearest 10 cm**?

a) 33 cm   b) 38 cm   c) 35 cm   d) 55 cm

3 a) 2 + 2 + 2 + 2 = ☐   b) ☐ × 2 = 8

c) 10 + 10 + 10 + 10 + 10 + 10 = ☐   d) ☐ × 10 = 60

4 Write down the missing **output** numbers.

a) input 4 → ×10 → ☐ output   b) input 5 → ×2 → ☐ output

c) input 7 → ×10 → ☐ output   d) input 8 → ×2 → ☐ output

5 a) $\frac{1}{2}$ of 18 = ☐   b) $\frac{1}{2}$ of 4 = ☐   c) $\frac{1}{2}$ of 8 = ☐

d) $\frac{1}{2}$ of 20 = ☐   e) $\frac{1}{2}$ of 14 = ☐   f) $\frac{1}{2}$ of 16 = ☐

g) $\frac{1}{2}$ of 12 = ☐   h) $\frac{1}{2}$ of 10 = ☐   i) $\frac{1}{2}$ of 2 = ☐

12 Number

Division by 2 and 10

Unit 12 words

equal	divide	share
left over	remainder	heaviest
lightest	odd	even
between	measure	length

Remember

Examples are shown in red.

means copy and complete.

 You need

- a set of Unit 12 vocabulary Snap cards.

 Play a game of Snap to help you learn the words.

 Try the **word test** to get some points.

1 2 friends share 10 buns equally.

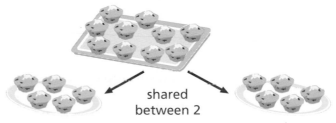

shared
between 2

They have 5 buns each.
10 buns **divided between 2** is **5** buns.

Divide these buns equally.

1 a) 8 buns divided between 2 people is ☐ buns.

b) 18 buns divided between 2 people is ☐ buns.

c) 14 buns divided between 2 people is ☐ buns.

2 Try Worksheet 1 *Equal shares (1)*.

3 Ro can jump **2 steps** at a time.

She starts on step 0.

How many jumps does she take to land on step **6**?

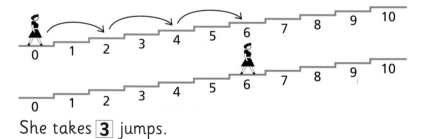

She takes **3** jumps.

How many jumps does Ro take:

3 a) to land on step **2**? She takes ☐ jump.

b) to land on step **8**? She takes ☐ jumps.

c) to land on step **10**? She takes ☐ jumps.

 4   Try Worksheet 2 *Jumping twos* and Worksheet 3 *Halve it*.

5   14p shared between 2 people is:

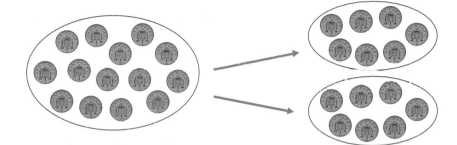

$14p \div 2 = 7p$

Remember

If we share between 2 people, we say we **divide by two** or **÷ 2**.

 5

a) $20p \div 2 = \square p$ b) $8p \div 2 = \square p$

c) $16p \div 2 = \square p$ d) $12p \div 2 = \square p$

 6   Try Worksheet 4 *Equal shares (2)*.

7   Here are different ways of showing **14 ÷ 2**:

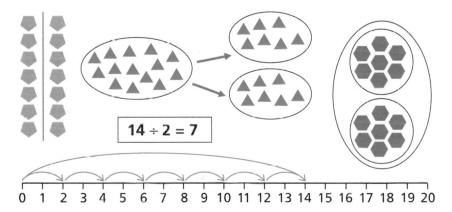

$14 \div 2 = 7$

Make your own chart for **18 ÷ 2**.

8 a) $8 \div 2 = \square$ b) $\square \div 2 = 4$

c) $14 \div 2 = \square$ d) $\square \div 2 = 7$

e) $18 \div 2 = \square$ f) $\square \div 2 = 9$

9 You want to make **2** teams from **8** people.
Pick two people at a time.

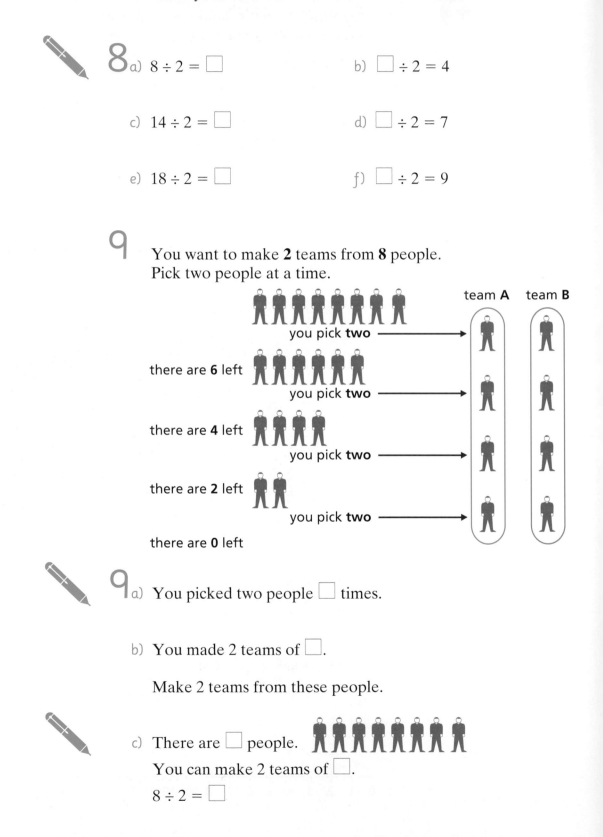

9 a) You picked two people $\square$ times.

b) You made 2 teams of $\square$.

Make 2 teams from these people.

c) There are $\square$ people.
You can make 2 teams of $\square$.
$8 \div 2 = \square$

d) There are ☐ people.
   You can make 2 teams of ☐.
   $10 \div 2 = $ ☐

e) There are ☐ people.
   You can make 2 teams of ☐.
   $14 \div 2 = $ ☐

 10   What would happen if you picked teams from 9 people?

11
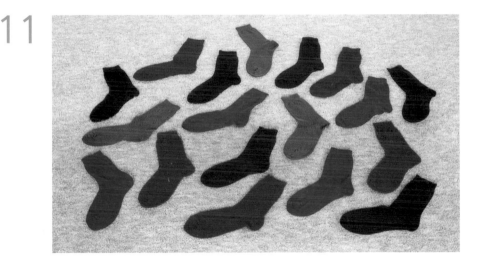

11 a) There are ☐ pairs of red socks.
      There is ☐ red sock left over.

   b) There are ☐ pairs of blue socks.
      There are ☐ blue socks left over.

   c) There is ☐ pair of grey socks.
      There is ☐ grey sock left over.

d) There are ☐ pairs of navy socks.

There are ☐ navy socks left over.

Complete each sentence with the word **odd** or **even**.

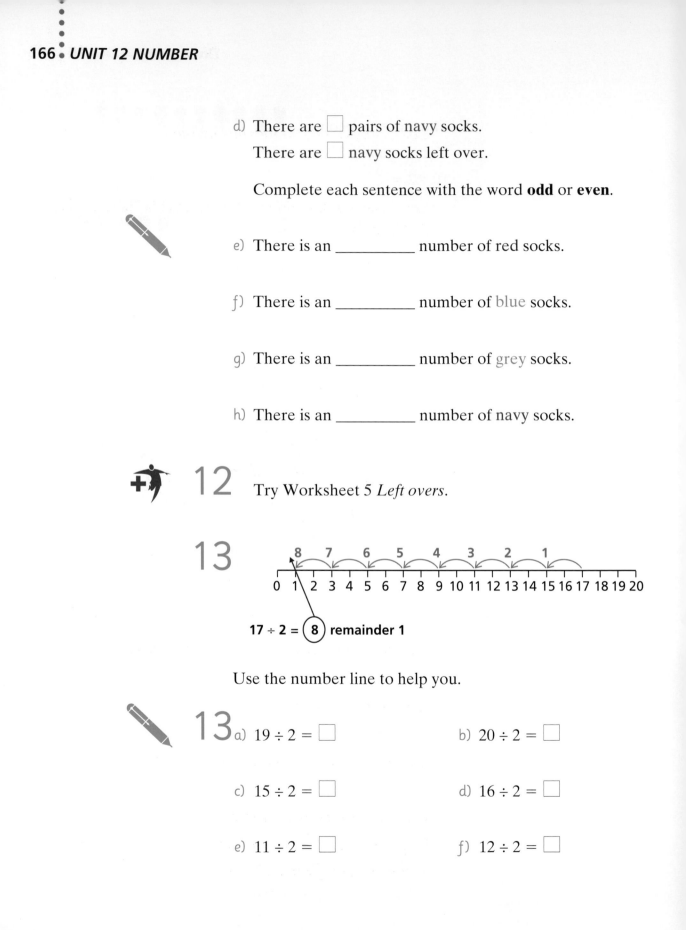

e) There is an _____ number of red socks.

f) There is an _____ number of blue socks.

g) There is an _____ number of grey socks.

h) There is an _____ number of navy socks.

12 Try Worksheet 5 *Left overs*.

13

$$17 \div 2 = \boxed{8} \text{ remainder } 1$$

Use the number line to help you.

13
a) $19 \div 2 = \square$

b) $20 \div 2 = \square$

c) $15 \div 2 = \square$

d) $16 \div 2 = \square$

e) $11 \div 2 = \square$

f) $12 \div 2 = \square$

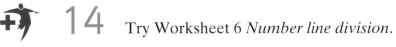 **14** Try Worksheet 6 *Number line division.*

15 Len clears out his cupboard. He gives his magazines to two friends.

He has **six** magazines.

He gives them to two friends: 6 ÷ 2 = 3

Each friend gets **3** magazines.

There are **0** magazines left over.

How many magazines would each friend get if:

15 a) Len has **ten** magazines?

Each friend gets ☐ magazines.

There are ☐ magazines left over.

b) Len has **seventeen** magazines?

Each friend gets ☐ magazines.

There is ☐ magazine left over.

c) Len has **eighteen** magazines?

Each friend gets ☐ magazines.

There are ☐ magazines left over.

d) Len has **thirteen** magazines?

Each friend gets ☐ magazines.

There is ☐ magazine left over.

e) Len has **fourteen** magazines?

Each friend gets ☐ magazines.

There are ☐ magazines left over.

f) Len has **fifteen** magazines?

Each friend gets ☐ magazines.

There is ☐ magazine left over.

 16 Make up some stories or word problems of your own for **15 divided by 2**.

17 10 friends share 30 buns equally.

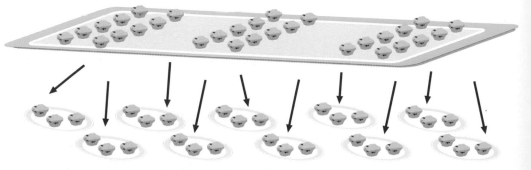

They have 3 buns each.
30 buns **divided between 10** is **3** buns.

Divide these trays of buns equally.

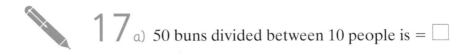 **17** a) 50 buns divided between 10 people is = ☐

b) 60 buns divided between 10 people is = ☐

c) 20 buns divided between 10 people is = ☐

d) 80 buns divided between 10 people is = ☐

18 Try Worksheet 7 *Ten shares*.

19 How many **jumps of ten** does it take to reach 70?

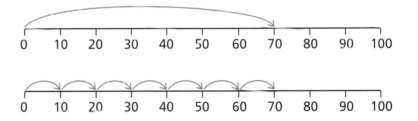

It takes **7** jumps of ten to reach 70.

19 a) It takes ☐ jumps of ten to reach 40.

b) It takes ☐ jumps of ten to reach 90.

c) It takes ☐ jumps of ten to reach 60.

20 Try Worksheet 8 *Jumping tens.*

21 Here are different ways of showing **30 ÷ 10**:

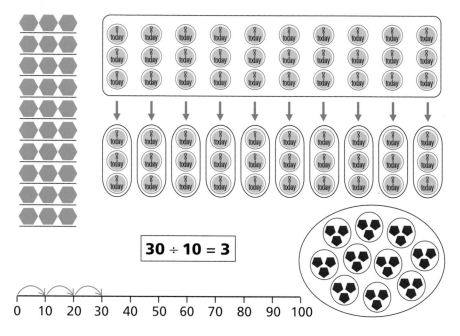

$$30 \div 10 = 3$$

Make your own chart for **40 ÷ 10**.

22 If you shared 26p among 10 people:

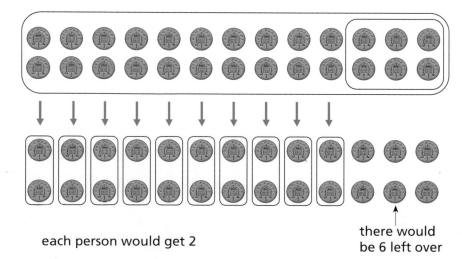

each person would get 2

there would
be 6 left over

Copy and complete the table, or use Worksheet 9.

	Each	Left over
20p ÷ 10	2p	0p
23p ÷ 10		
93p ÷ 10		
10p ÷ 10		
71p ÷ 10		
40p ÷ 10		
55p ÷ 10		
80p ÷ 10		
30p ÷ 10		

23 Pat likes gardening.

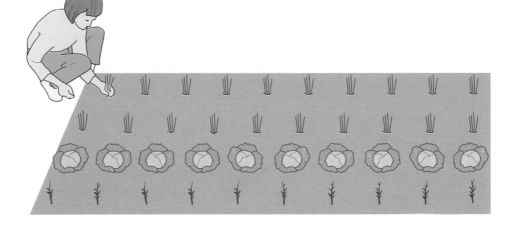

She plants all her vegetables in rows of ten.

She can plant **sixty** cabbages in 6 rows with
 0 cabbages left over.

She can plant:

23 a) **thirty** in ☐ rows with ☐ left over

b) **thirty-two** in ☐ rows with ☐ left over

c) **fifty** in ☐ rows with ☐ left over

d) **fifty-eight** in ☐ rows with ☐ left over

e) **seventy** in ☐ rows with ☐ left over

f) **seventy-one** in ☐ rows with ☐ left over

g) **eighty** in ☐ rows with ☐ left over

h) **eighty-eight** in ☐ rows with ☐ left over

i) **one hundred** in ☐ rows with ☐ left over.

24 You need

- Unit 12 Race against time cards
- Unit 5 Race against time cards
- your 'My maths record' sheet.

Race against time

1 Sort the Unit 12 race cards – this side up. | $20 \div 10$ |

2 Take the cards 1 at a time.

3 Answer as quickly as you can.

4 Did you get it right?

Look at the other side of the card. | 2 |

5 When you get all the answers correct, ask a friend to test you.

6 Now **Race against time**

Go for points!

Ask your teacher to test and time you.

7 Now try the Unit 5 race cards – this side up. | 2×10 |

8 You can go for more points.

Remember

3 errors – 1 point

2 errors – 2 points

1 error – 3 points

0 errors – 5 points

Answer in 1 minute with 0 errors – 7 points

Now try Unit 12 Test.

Review 12

1 a) My **1 m reminder** is _____.

   b) My **1 kg reminder** is _____.

2 Would you use a **metre** measure or a **kilogram** measure to find the length of a table?

   I would use a _____ measure.

3 How much change would you have from 20p if something cost:

3 a) 17p?   change ☐p   b) 12p?   change ☐p

   c) 5p?   change ☐p   d) 9p?   change ☐p

   e) 12p?   change ☐p   f) 14p?   change ☐p

4 Write these numbers in order, **smallest** number first.

413   14   43   134   341   4

5 a) $7 + 9 = $ ☐   b) $14 + 6 = $ ☐

   c) $7 + 7 = $ ☐   d) $3 + 9 = $ ☐

e) $14 + 5 = \square$ f) $5 + 11 = \square$

g) $8 + 4 = \square$ h) $4 + 8 = \square$

i) $6 + 13 = \square$ j) $11 + 9 = \square$

k) $7 + 13 = \square$ l) $15 + 4 = \square$

6   What is the **difference** between these ages:

6 a) twenty years old and six years old? _____ years

b) seventeen years old and ten years old? _____ years

c) seventeen years old and nine years old? _____ years

7   How are these two shapes alike?

13 Measures

Estimating and measuring volume

Unit 13 words

litre	half litre	divide
share	left over	remainder
gram	most	fewer
measure	block graph	multiply

Remember

Examples are shown in red.

 means copy and complete.

 You need

- a set of Unit 13 vocabulary Snap cards.

 Play a game of Snap to help you learn the words.

Try the **word test** to get some points.

1 Some friends buy the same drink in these glasses.
They each pay **80p** for **one** glass.

1 a) Was each friend charged a fair price?

b) A recipe says 'add two cups of water'.
Tell your partner why you think this could be a problem.

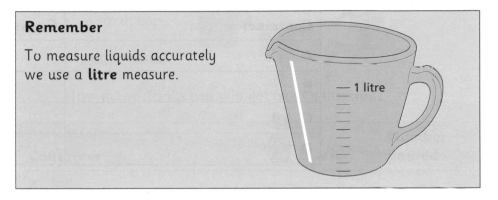

Remember

To measure liquids accurately
we use a **litre** measure.

— 1 litre

2 You need

- a litre jug

- an egg cup

- a spoon

- a yoghurt pot

- a mug

- some water.

and

2 a) Copy the table on the next page, or use Worksheet 1.

Fill in the **guess** column before you begin.

b) Use each container in turn to fill the half-litre jug with water.

Copy and complete the table.

Container		Guess	Number needed to fill half-litre jug
A			
B			
C			

c) Which container holds about **half a litre ($\frac{1}{2}$ litre)**?

_____ holds about half a litre.

> **Remember**
>
> Remember this container as your own **half-litre reminder**.

6 Try Worksheet 4 *Colour half a litre*.

7 You need

- a litre jug

- a half-litre jug

- 2 empty milk bottles or cartons

- some water.

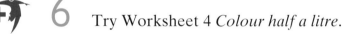

and

We often buy milk in **pint** bottles.

7 a) Find out **roughly** how many pints make $\frac{1}{2}$ litre.

There is roughly ☐ pint in $\frac{1}{2}$ litre.

b) Find out **about** how many pints make 1 litre.

There are about ☐ pints in 1 litre.

c) There are ☐ $\frac{1}{2}$-litre measures in 1 litre.

8 You need

- a half-litre jug  and
- some water.

You find a container that you think would hold half a litre.

Your partner finds a container which he or she thinks would hold half a litre.

Use your half-litre jug to see who is correct.

1 point for reach correct measure. (It could be a draw!)

> **Remember**
>
> One litre is often written as 1 l.
>
> Half a litre is often written as $\frac{1}{2}$ l.

9 Match the picture to the measure.

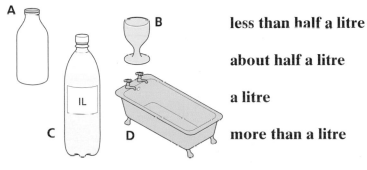

less than half a litre

about half a litre

a litre

more than a litre

9 a) **A** holds <u>**about half a litre**</u>.

b) **B** holds _____ .

c) **C** holds _____ .

d) **D** holds _____ .

10 Ice-cream is sold in litres.

Strawberry
$\frac{1}{2}$ litre

Mint
1 litre

Chocolate
2 litres

Vanilla
4 litres

10 a) 1 tub of strawberry

   holds ☐ litre.

b) 2 tubs of strawberry

   hold ☐ litre.

c) 1 tub of vanilla

   holds ☐ litres.

d) 2 tubs of vanilla

   hold ☐ litres.

e) 1 tub of chocolate

holds ☐ litres.

f) 10 tubs of chocolate

hold ☐ litres.

g) 1 tub of vanilla holds ☐ **more** litres than a tub of mint.

h) 1 tub of mint holds ☐ **fewer** litres than a tub of vanilla.

i) ☐ tubs of strawberry hold **the same** as a tub of mint.

 11 Make a list of other things that are sold in litres.

12 A group of people were asked how much they drink each day. This tally table shows the results.

Amount drunk	Tally marks
$\frac{1}{2}$ litre	I
1 litre	︍HHH IIII
2 litres	︍HHH II
3 litres	III

 Draw a **block graph** to show these results.

You need

• squared paper or Worksheet 5.

Remember

Remember the **rules for drawing a block graph** on page 22.

Class guess

13 Play a game of 'Class guess' each day for a week.

Rules for 'Class guess'

1 Somebody holds up a container such as a bucket or a cup, or a picture of a container.
2 Everyone tries to guess how many litres it would take to fill it.
3 Everyone writes down their guesses.
4 Two people collect up the written guesses.
5 Someone fills the container with water (where possible) and counts how many litres it takes.
6 Your teacher will tell you how near the guess has to be to be a 'good' guess.
7 Keep score of the 'good' guesses.

Monday – **1** 'good' guess.

8 At the end of the week, draw a **block graph**.

Put 5 columns along the bottom, for Monday to Friday.
Put the numbers for 'good' guesses up the side.

Are there more 'good' guesses as the week goes on?

Now try Unit 13 Test.

Review 13

1 My **1 kg reminder** is _____.

2 My **1 m reminder** is _____.

3 Would you use a **metre** measure or a **kilogram** measure to find your weight?

I would use a _____ measure.

4 a) $10 + 8 = \square$

   $18 - \square = 10$

   $8 + 10 = \square$

   $18 - \square = 8$

   b) $10 + 4 = \square$

   $14 - \square = 10$

   $4 + 10 = \square$

   $14 - \square = 4$

5 a) $5 \times 2 = \square$

   b) $7 \times 10 = \square$

   c) $5 \times 10 = \square$

   d) $8 \times 2 = \square$

   e) $6 \times 10 = \square$

   f) $3 \times 2 = \square$

   g) $4 \times 10 = \square$

   h) $2 \times 2 = \square$

6 Write these numbers in order, **smallest** number first.

   231   103   213   312   123   301

7   Which shapes are **symmetrical**?

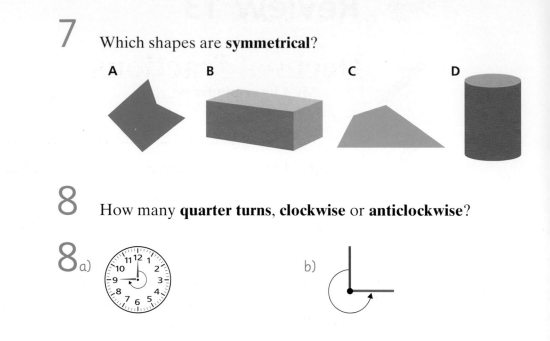

A   B   C   D

8   How many **quarter turns**, **clockwise** or **anticlockwise**?

8 a)   b)

14 Number

Decimal fractions

Unit 14 words

eighty	ninety	whole
tenth	decimal	divide
half	quarter	equal
litre	half litre	hundred

Remember

Examples are shown in red.

means copy and complete.

You need

- a set of Unit 14 vocabulary Snap cards.

Play a game of Snap to help you learn the words.

Try the **word test** to get some points.

Remember

If we divide something into **2 equal parts**, each part is a **half** ($\frac{1}{2}$).

1

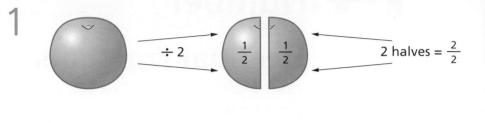

$\div\,2$ 2 halves $= \dfrac{2}{2}$

1 a) The orange is cut into ☐ equal parts.

b) Each part is called a _____.

2 Try Worksheet 1 *Half remembered*.

Remember

If we divide something into **4 equal parts**, each part is a
quarter ($\frac{1}{4}$).

3

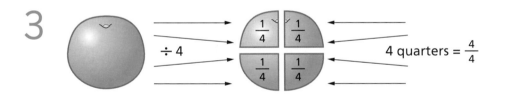

$\div\,4$ 4 quarters $= \dfrac{4}{4}$

3 a) The orange is cut into ☐ equal parts.

b) Each part is called a _____.

4 Try Worksheet 2 *Quarter remembered*.

> **Remember**
>
> If we divide something into **10 equal parts**, each part is a **tenth** ($\frac{1}{10}$).

5

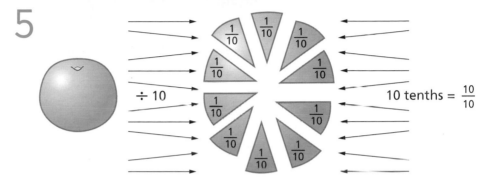

$\div\ 10$

10 tenths = $\frac{10}{10}$

5 a) The orange is cut into ☐ equal parts.

    b) Each part is called a _____.

6   Try Worksheet 3 *Tenths*.

7   Pat and Rani share a bar of chocolate.

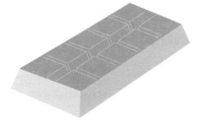

Pat divides it into ten chunks.

Each chunk is **one tenth** ($\frac{1}{10}$) of the bar.

Rani takes **two** chunks.

She takes **two** tenths: $\boxed{\tfrac{1}{10}}\boxed{\tfrac{1}{10}} = \dfrac{\mathbf{2}}{10}$

Pat has **eight** tenths left: $\boxed{\tfrac{1}{10}}\boxed{\tfrac{1}{10}}\boxed{\tfrac{1}{10}}\boxed{\tfrac{1}{10}} = \dfrac{\mathbf{8}}{10}$
$\boxed{\tfrac{1}{10}}\boxed{\tfrac{1}{10}}\boxed{\tfrac{1}{10}}\boxed{\tfrac{1}{10}}$

7 a) Rani takes **four** chunks.

She takes _____ tenths: $\boxed{\tfrac{1}{10}}\boxed{\tfrac{1}{10}}\boxed{\tfrac{1}{10}}\boxed{\tfrac{1}{10}} = \dfrac{\square}{10}$

Pat has _____ tenths left: $\boxed{\tfrac{1}{10}}\boxed{\tfrac{1}{10}}\boxed{\tfrac{1}{10}}\boxed{\tfrac{1}{10}}\boxed{\tfrac{1}{10}}\boxed{\tfrac{1}{10}} = \dfrac{\square}{10}$

b) Rani takes **six** chunks.

She takes _____ tenths: $\boxed{\tfrac{1}{10}}\boxed{\tfrac{1}{10}}\boxed{\tfrac{1}{10}}\boxed{\tfrac{1}{10}}\boxed{\tfrac{1}{10}}\boxed{\tfrac{1}{10}} = \dfrac{\square}{10}$

Pat has _____ tenths left: $\boxed{\tfrac{1}{10}}\boxed{\tfrac{1}{10}}\boxed{\tfrac{1}{10}}\boxed{\tfrac{1}{10}} = \dfrac{\square}{10}$

c) Rani takes **two** chunks.

She takes _____ tenths: $\boxed{\tfrac{1}{10}}\boxed{\tfrac{1}{10}} = \dfrac{\square}{10}$

Pat has _____ tenths left: $\boxed{\tfrac{1}{10}}\boxed{\tfrac{1}{10}}\boxed{\tfrac{1}{10}}\boxed{\tfrac{1}{10}}\boxed{\tfrac{1}{10}}\boxed{\tfrac{1}{10}}\boxed{\tfrac{1}{10}}\boxed{\tfrac{1}{10}} = \dfrac{\square}{10}$

d) Rani takes **ten** chunks.

She takes _____ tenths: $\boxed{\frac{1}{10}}\boxed{\frac{1}{10}}\boxed{\frac{1}{10}}\boxed{\frac{1}{10}}\boxed{\frac{1}{10}}$ = $\dfrac{\square}{10}$

$\boxed{\frac{1}{10}}\boxed{\frac{1}{10}}\boxed{\frac{1}{10}}\boxed{\frac{1}{10}}\boxed{\frac{1}{10}}$

Pat has _____ tenths left: $\dfrac{\square}{10}$

8 a) How many different ways can **2 people** divide a bar of **ten chunks** of chocolate?

b) How many tenths make the whole bar?

9 Try Worksheet 4 *Whole one*.

10 Jo has a different bar of chocolate.

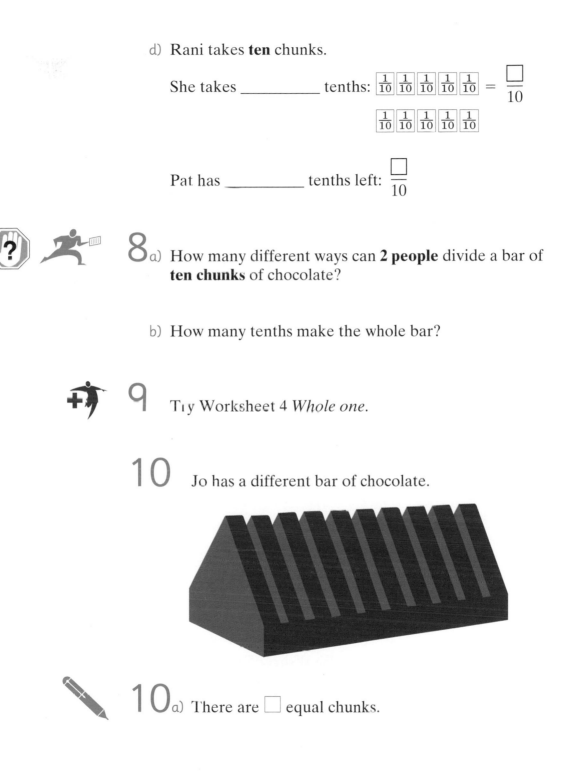

10 a) There are $\square$ equal chunks.

b) Each chunk is ☐ tenth.

Jo does not eat the whole bar.

He eats **one** out of **ten** chunks.

He eats **1 tenth**, $\frac{1}{10}$.

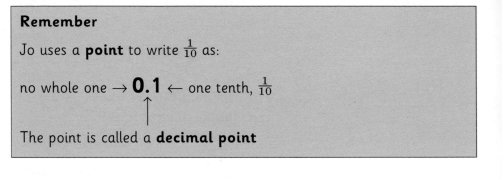

Remember

Jo uses a **point** to write $\frac{1}{10}$ as:

no whole one → **0.1** ← one tenth, $\frac{1}{10}$

The point is called a **decimal point**

c) Jo eats ☐ whole bars.

nought point one

d) Jo eats ☐ tenth of a bar.

e) Jo writes it as ☐.☐.

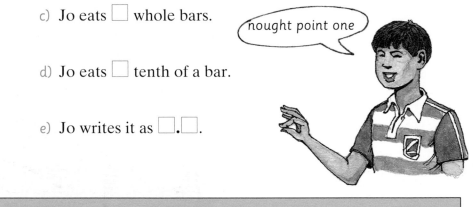

Remember

A number with a **decimal point** is called a **decimal number**.

11   Try Worksheet 5 *Decimal numbers*.

12 Look at the ruler.

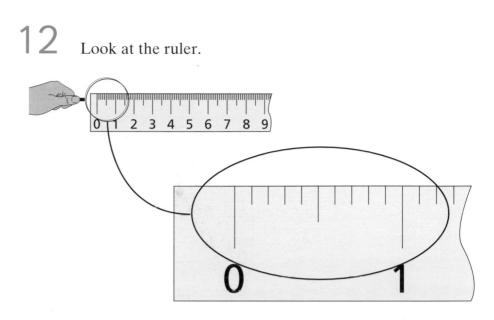

If we looked through a magnifying glass at 1 cm we would see that each centimetre is divided into **10 equal parts**.

Each part is **one-tenth ($\frac{1}{10}$)** of a centimetre.

Copy and complete, or use Worksheet 6.

12

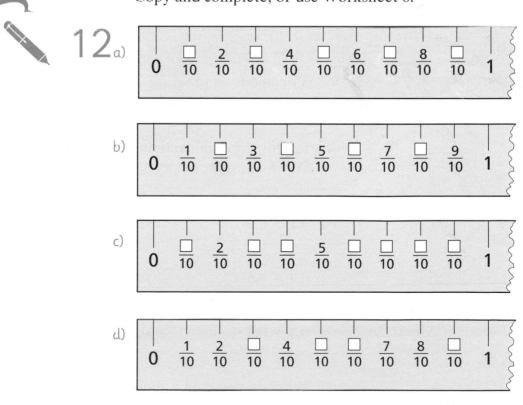

a) $0 \quad \square \quad \frac{2}{10} \quad \square \quad \frac{4}{10} \quad \square \quad \frac{6}{10} \quad \square \quad \frac{8}{10} \quad \square \quad 1$

b) $0 \quad \frac{1}{10} \quad \square \quad \frac{3}{10} \quad \square \quad \frac{5}{10} \quad \square \quad \frac{7}{10} \quad \square \quad \frac{9}{10} \quad 1$

c) $0 \quad \square \quad \frac{2}{10} \quad \square \quad \square \quad \frac{5}{10} \quad \square \quad \square \quad \square \quad 1$

d) $0 \quad \frac{1}{10} \quad \frac{2}{10} \quad \square \quad \frac{4}{10} \quad \square \quad \square \quad \frac{7}{10} \quad \frac{8}{10} \quad \square \quad 1$

13 Jo fills in the missing numbers as **decimals**.

Copy and complete, or use Worksheet 6.

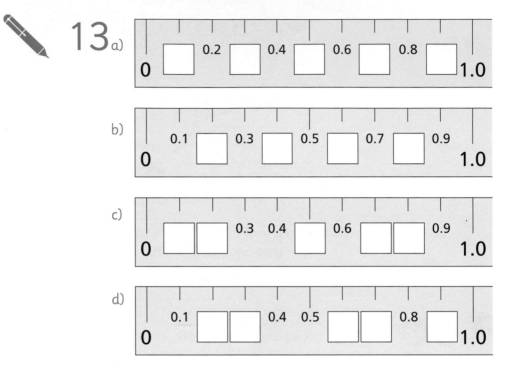

13 a) 0 [ ] 0.2 [ ] 0.4 [ ] 0.6 [ ] 0.8 [ ] 1.0

b) 0 0.1 [ ] 0.3 [ ] 0.5 [ ] 0.7 [ ] 0.9 1.0

c) 0 [ ] [ ] 0.3 0.4 [ ] 0.6 [ ] [ ] 0.9 1.0

d) 0 0.1 [ ] [ ] 0.4 0.5 [ ] [ ] 0.8 [ ] 1.0

14 You need

• a calculator.

Put these **decimal numbers** into the calculator.
Write down the answers.

14 a) 0.1 + 0.9   b) 0.2 + 0.8   c) 0.3 + 0.7

d) 0.4 + 0.6   e) 0.5 + 0.5   f) 0.6 + 0.4

g) 0.7 + 0.3   h) 0.8 + 0.2   i) 0.9 + 0.1

j) What do you notice about all the answers?

15 Gita has two bars of chocolate.

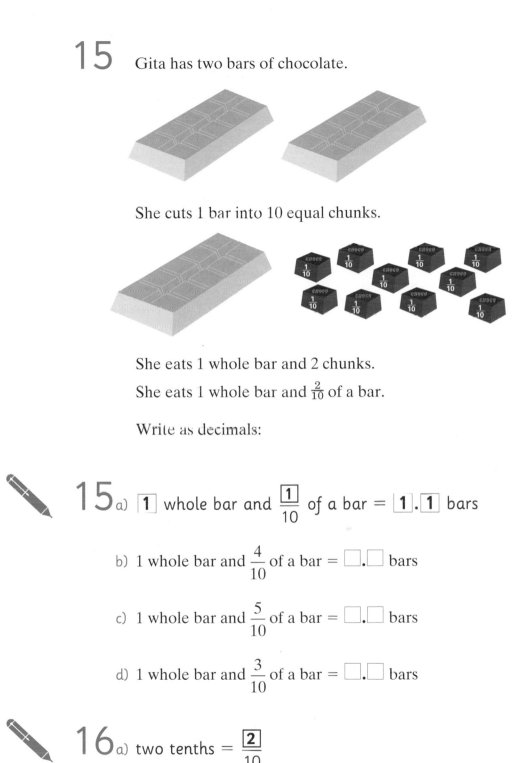

She cuts 1 bar into 10 equal chunks.

She eats 1 whole bar and 2 chunks.

She eats 1 whole bar and $\frac{2}{10}$ of a bar.

Write as decimals:

15 a) $\boxed{1}$ whole bar and $\frac{\boxed{1}}{10}$ of a bar = $\boxed{1}.\boxed{1}$ bars

b) 1 whole bar and $\frac{4}{10}$ of a bar = $\square.\square$ bars

c) 1 whole bar and $\frac{5}{10}$ of a bar = $\square.\square$ bars

d) 1 whole bar and $\frac{3}{10}$ of a bar = $\square.\square$ bars

16 a) two tenths = $\frac{\boxed{2}}{10}$

b) one whole one and two tenths = $\boxed{1}\ \frac{\boxed{2}}{10}$

c) six tenths = $\dfrac{\square}{10}$

d) one whole one and six tenths = $\square\dfrac{\square}{10}$

e) nine tenths = $\dfrac{\square}{10}$

f) one whole one and nine tenths = $\square\dfrac{\square}{10}$

17 Try Worksheet 7 *More decimal numbers.*

18 a) $\dfrac{\boxed{2}}{10} = \boxed{0}.\boxed{2}$ b) $\boxed{1}\dfrac{\boxed{2}}{10} = \boxed{1}.\boxed{2}$

c) $1\dfrac{3}{10} = \square.\square$ d) $\dfrac{3}{10} = \square.\square$

e) $\dfrac{6}{10} = \square.\square$ f) $1\dfrac{6}{10} = \square.\square$

g) $\dfrac{8}{10} = \square.\square$ h) $1\dfrac{8}{10} = \square.\square$

i) $1 = \square.\square$

19 Write as fractions.

19 a) $1.4 = \boxed{1}\dfrac{\boxed{4}}{10}$ b) $0.4 = \dfrac{\boxed{4}}{10}$

c) $1.9 = \square\dfrac{\square}{10}$ d) $0.9 = \dfrac{\square}{10}$

e) $1.1 = \Box \dfrac{\Box}{10}$

f) $0.1 = \dfrac{\Box}{10}$

g) $1.6 = \Box \dfrac{\Box}{10}$

h) $0.6 = \dfrac{\Box}{10}$

i) $1.2 = \Box \dfrac{\Box}{10}$

j) $0.2 = \dfrac{\Box}{10}$

k) $1.5 = \Box \dfrac{\Box}{10}$

l) $0.5 = \dfrac{\Box}{10}$

m) $1.8 = \Box \dfrac{\Box}{10}$

n) $0.8 = \dfrac{\Box}{10}$

20 Try Worksheets 8 and 9 *Shading decimals (1) and (2)*.

21 Look at the ruler.

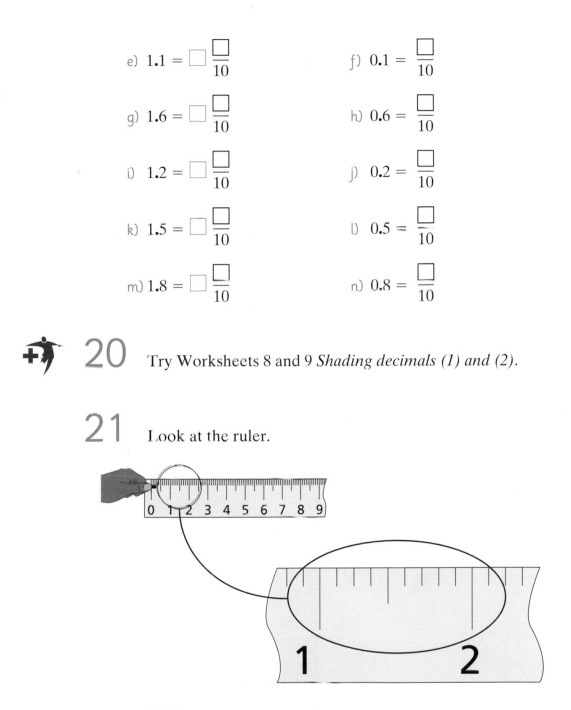

If we looked through a magnifying glass at 1–2 cm we would see that each centimetre is divided into **10 equal parts**.

Each part is **one tenth ($\frac{1}{10}$)** of a centimetre.

Copy and complete, or use Worksheet 10.

21 a)

| 1 | $1\frac{\square}{10}$ | $1\frac{2}{10}$ | $1\frac{\square}{10}$ | $1\frac{\square}{10}$ | $1\frac{5}{10}$ | $1\frac{\square}{10}$ | $1\frac{\square}{10}$ | $1\frac{\square}{10}$ | $1\frac{\square}{10}$ | 2 |

b)

| 1 | $1\frac{1}{10}$ | $1\frac{2}{10}$ | $1\frac{3}{10}$ | $1\frac{\square}{10}$ | $1\frac{\square}{10}$ | $1\frac{\square}{10}$ | $1\frac{7}{10}$ | $1\frac{\square}{10}$ | $1\frac{9}{10}$ | 2 |

c)

| 1 | $1\frac{1}{10}$ | $1\frac{\square}{10}$ | $1\frac{3}{10}$ | $1\frac{4}{10}$ | $1\frac{\square}{10}$ | $1\frac{6}{10}$ | $1\frac{\square}{10}$ | $1\frac{8}{10}$ | $1\frac{\square}{10}$ | 2 |

d)

| 1 | $1\frac{\square}{10}$ | $1\frac{2}{10}$ | $1\frac{\square}{10}$ | $1\frac{4}{10}$ | $1\frac{\square}{10}$ | $1\frac{6}{10}$ | $1\frac{\square}{10}$ | $1\frac{8}{10}$ | $1\frac{\square}{10}$ | 2 |

22 Jo fills in the missing numbers as decimals.

Copy and complete, or use Worksheet 10.

22 a)

| 1 | □ | 1.2 | □ | 1.4 | □ | 1.6 | □ | 1.8 | □ | 2 |

b)

| 1 | 1.1 | □ | 1.3 | □ | 1.5 | □ | 1.7 | □ | 1.9 | 2 |

c)

| 1 | □ | □ | 1.3 | 1.4 | □ | □ | 1.7 | 1.8 | □ | 2 |

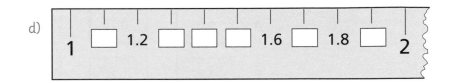

d)

| 1 | ☐ | 1.2 | ☐ | ☐ | ☐ | 1.6 | ☐ | 1.8 | ☐ | 2 |

23 You need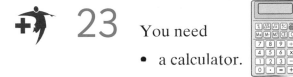

- a calculator.

Put these **decimal numbers** into the calculator.
Write down the answers.

23 a) 1.1 + 0.9 b) 1.2 + 0.8 c) 1.3 + 0.7

d) 1.4 + 0.6 e) 1.5 + 0.5 f) 1.6 + 0.4

g) 1.7 + 0.3 h) 1.8 + 0.2 i) 1.9 + 0.1

j) What do you notice about all the answers?

Speed sort

24 You need

- one set of 'Speed sort' cards each. (Do not mix the sets up.)

Play a game of 'Speed sort'.

Rules for 'Speed sort'

1 Both shuffle your set of cards.
2 Look at line **A** on the next page.
3 See how quickly you and your partner can:
  a) find the cards with the same numbers
  b) put them in order, **smallest** number first.
4 The first player with a set of cards in order is the winner.
5 Then try line **B**, then line **C**, then line **D**.

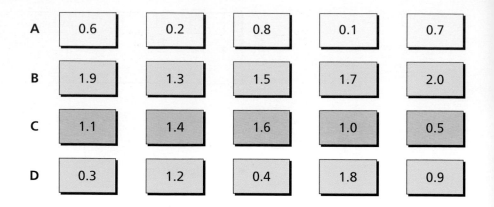

A	0.6	0.2	0.8	0.1	0.7
B	1.9	1.3	1.5	1.7	2.0
C	1.1	1.4	1.6	1.0	0.5
D	0.3	1.2	0.4	1.8	0.9

 **25**   Try Worksheet Puzzle *Different colours.*

26   You need

- Unit 14 Race against time cards
- your 'My maths record' sheet.

Race against time

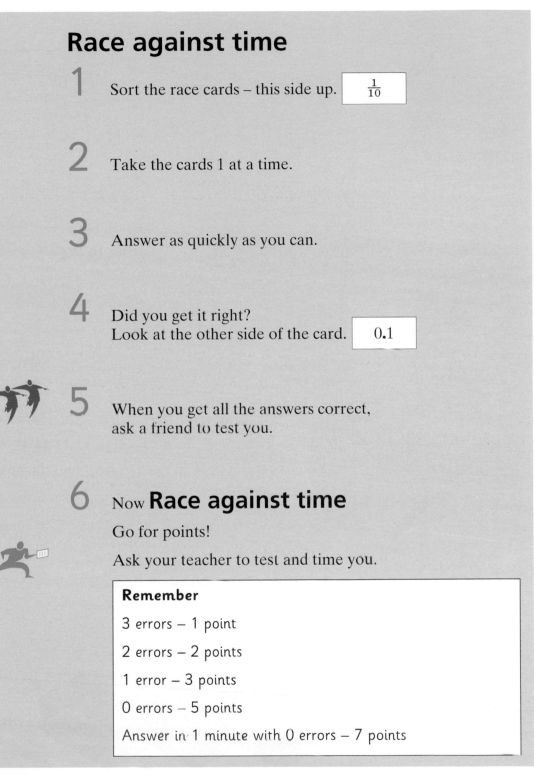

1   Sort the race cards – this side up. $\frac{1}{10}$

2   Take the cards 1 at a time.

3   Answer as quickly as you can.

4   Did you get it right?
Look at the other side of the card.   0.1

5   When you get all the answers correct,
ask a friend to test you.

6   Now **Race against time**

Go for points!

Ask your teacher to test and time you.

> **Remember**
>
> 3 errors – 1 point
>
> 2 errors – 2 points
>
> 1 error – 3 points
>
> 0 errors – 5 points
>
> Answer in 1 minute with 0 errors – 7 points

Now try Unit 14 Test.

Review 14

1
a) $6 \div 2 = \square$

b) $80 \div 10 = \square$

c) $30 \div 10 = \square$

d) $6 \div 2 = \square$

e) $40 \div 10 = \square$

f) $8 \div 2 = \square$

g) $60 \div 10 = \square$

h) $14 \div 2 = \square$

i) $70 \div 10 = \square$

j) $2 \div 2 = \square$

k) $90 \div 10 = \square$

l) $4 \div 2 = \square$

2

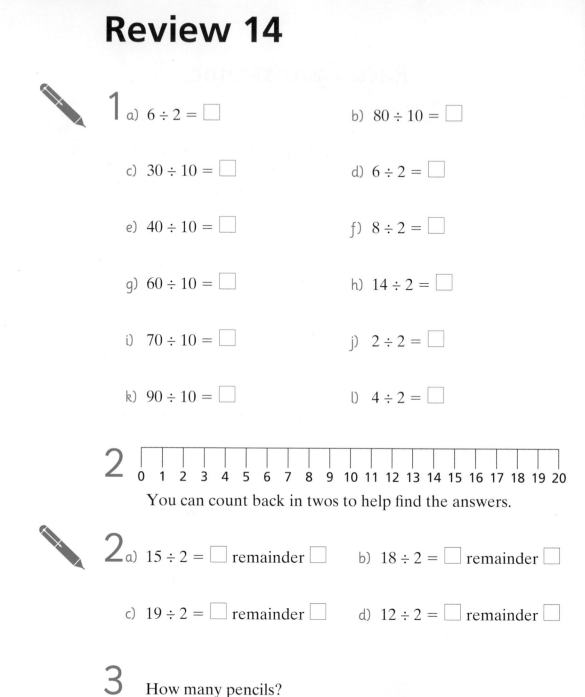

0 1 2 3 4 5 6 7 8 9 10 11 12 13 14 15 16 17 18 19 20

You can count back in twos to help find the answers.

2
a) $15 \div 2 = \square$ remainder $\square$

b) $18 \div 2 = \square$ remainder $\square$

c) $19 \div 2 = \square$ remainder $\square$

d) $12 \div 2 = \square$ remainder $\square$

3 How many pencils?

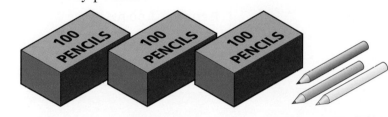

There are $\square$ pencils altogether.

4 My **half-litre reminder** is _____.

5 Match the picture to the measure.

less than $\frac{1}{2}$ litre

about $\frac{1}{2}$ litre

about 1 litre

more than 1 litre

A – about $\frac{1}{2}$ litre

6 Name 2 things you could buy in **litres**.

15 Measures

Time

Unit 15 words

time	minute	second
hour	whole	tenth
decimal	plus	subtract
multiply	divide	remainder

Remember

Examples are shown in red.

 means copy and complete.

 You need

- a set of Unit 15 vocabulary Snap cards.

 Play a game of Snap to help you learn the words.

 Try the **word test** to get some points.

1   Bob did these jobs.

A   B   C   D

1 a) Job _____ took the **longest time**.

b) Job _____ took the **shortest time**.

c) List the jobs in order, with the job that took the **shortest** time first.

2 You need

- a stop-watch.

2 a) Choose something you can talk about easily, such as your favourite sport, or last weekend.

b) Talk for 1 minute.

c) The person who keeps talking with fewest pauses is the winner.

Long ago, photographers took photographs like this.

They had no second hand on their watches.

They had to count to 30 seconds to take the photographs.

Some photographers used cameras called 'Kodak' cameras.

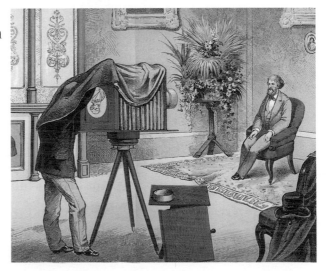

To time 30 seconds, photographers used to say:

one 'Kodak'
two 'Kodak'
three 'Kodak'

They counted up to

thirty 'Kodak'.

 3 a) You need

• a stop-watch.

Imagine you are taking a photograph. See how well you can count to **30 seconds**.

Make up a word of your own to help you get it right.

Ask a friend to time you.

b) Now try to do the same with **1 minute**.

Remember

60 seconds = 1 minute

 4 You need

• a stop-watch.

How many of the things in the table on the next page can you do in **1 minute**?

Ask a friend to time you.

Activity	Number of times
lift your arm	
sharpen a pencil	
tidy your desk	
write your full name	
put on a shoe	

Remember

60 minutes = 1 hour

5 How long would it take to do these things?

Choose **more than 1 hour** or **less than 1 hour**.

5 a) You would cook a large chicken

in _____.

b) You would boil an egg

in _____.

c) You would play a game of football

in _____.

d) You would run a 100 m sprint

in _____.

e) Now make a list of things that would take you **about 1 hour** to do. For example, watch a TV programme, bake a cake, do an exam …

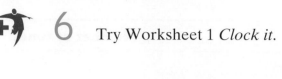

6 Try Worksheet 1 *Clock it*.

7 Mo turns **clockwise** from
here   to   here.

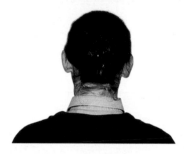 

Copy and complete, using **quarter ($\frac{1}{4}$)**, or **half ($\frac{1}{2}$)**, or

three-quarters ($\frac{3}{4}$).

She made a _____ $\frac{\Box}{\Box}$ turn.

8 The big hand on the clock moves from
here   to   here.

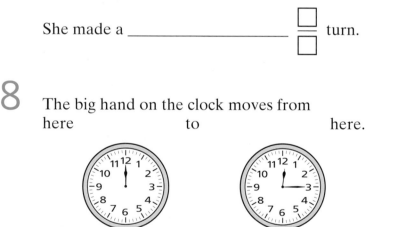

8 a) Copy and complete, using **($\frac{1}{4}$)**, or **half ($\frac{1}{2}$)**, or

three-quarters ($\frac{3}{4}$).

The big hand has made a _____ $\frac{\Box}{\Box}$ turn.

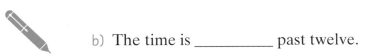 b) The time is _____ past twelve.

c) The big hand has moved ☐ minutes.

Remember

We can write **quarter past twelve** as **12:15**.

d) Here is a digital clock:

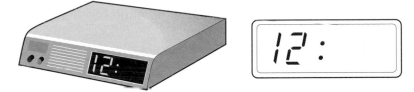

Draw the clock panel and show the time as **quarter past twelve**.

9 Mo turns **clockwise** from
here   to   here.

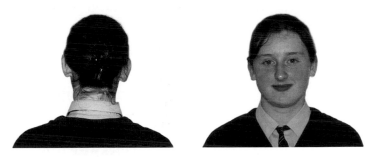

Copy and complete, using (¼), or **half** (½), or

three-quarters (¾).

She made a _____ $\frac{\square}{\square}$ turn.

10 The big hand on the clock moves from
here   to   here.

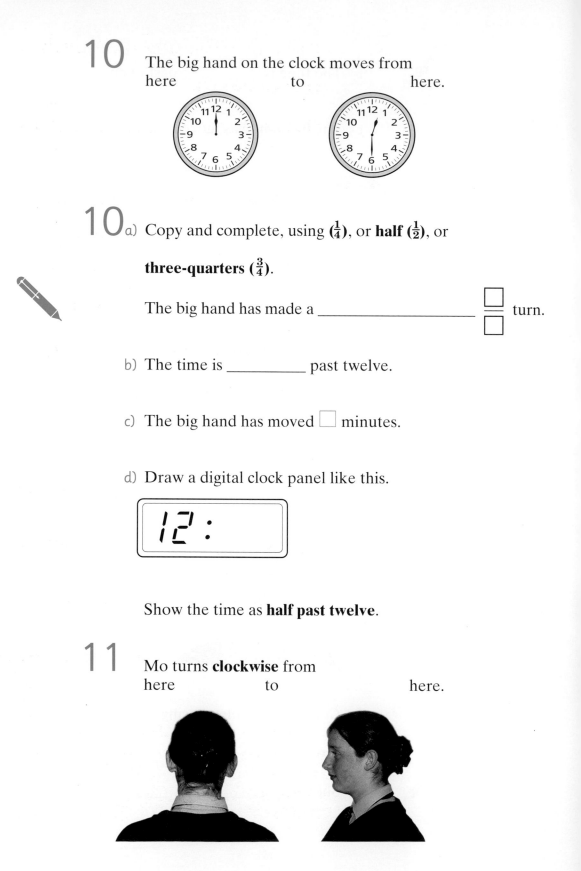

10 a) Copy and complete, using ($\frac{1}{4}$), or **half** ($\frac{1}{2}$), or

three-quarters ($\frac{3}{4}$).

The big hand has made a _____ ⬜/⬜ turn.

b) The time is _____ past twelve.

c) The big hand has moved ⬜ minutes.

d) Draw a digital clock panel like this.

12:

Show the time as **half past twelve**.

11 Mo turns **clockwise** from
here   to   here.

Copy and complete, using (¼), or **half** (½), or

three-quarters (¾).

She made a _____ $\dfrac{\square}{\square}$ turn.

12 The big hand on the clock moves from
here   to   here.

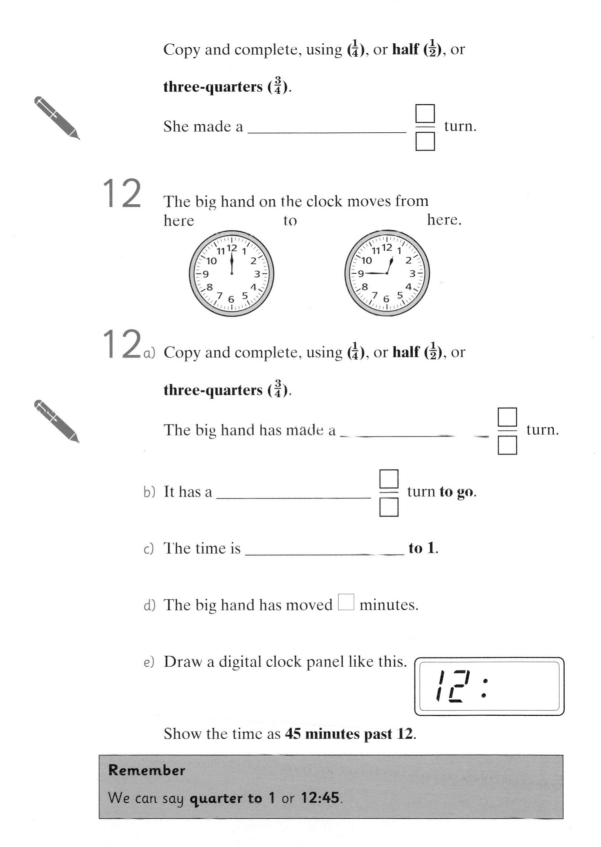

12 a) Copy and complete, using (¼), or **half** (½), or

three-quarters (¾).

The big hand has made a _____ $\dfrac{\square}{\square}$ turn.

b) It has a _____ $\dfrac{\square}{\square}$ turn **to go**.

c) The time is _____ **to 1**.

d) The big hand has moved ☐ minutes.

e) Draw a digital clock panel like this.

$$12:$$

Show the time as **45 minutes past 12**.

Remember

We can say **quarter to 1** or **12:45**.

13 Try Worksheet 2 *Which quarter?*

14 This clock shows the 60 minutes around a clock face.

> **Remember**
>
> There are 60 minutes in 1 hour.

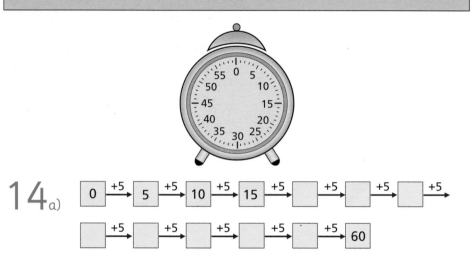

14 a)

| 0 | +5 | 5 | +5 | 10 | +5 | 15 | +5 | ☐ | +5 | ☐ | +5 | ☐ | +5 |

| ☐ | +5 | ☐ | +5 | ☐ | +5 | ☐ | +5 | ☐ | +5 | 60 |

b) In this clock, the big hand has moved from
here   to   here.

c) The big hand has moved ☐ minutes.

d) The clock shows ☐ minutes **past twelve**.

15 Try Worksheet 3 *Five minutes*.

16 Sue made a poster for an evening sponsored walk.
She left something out.

> **Remember**
>
> **a.m.** means **morning**.
>
> **p.m.** means **afternoon** or **evening**.

> **SPONSORED WALK**
>
> leaves
>
> Town Hall
>
> Saturday
> 1st September
>
> at 2.00

morning or evening?

Sue should have written 2.00 _____ .

17 Look at the time line below.

> **Remember**
>
> 24 hours = 1 day

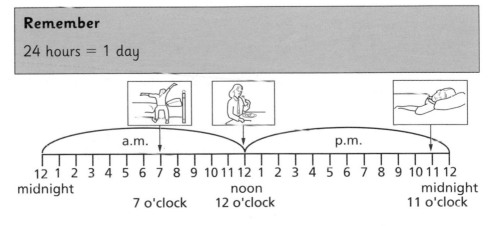

Copy and complete, using **a.m.** or **p.m.**

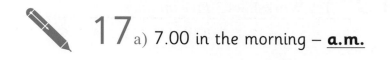

17

a) 7.00 in the morning – **a.m.**

b) 11.00 in the evening – _____

c) 3.00 in the afternoon – _____

d) 5.00 in the morning – _____

e) 4.00 in the morning – _____

f) 2.00 in the morning – _____

g) 2.00 in the afternoon – _____

h) 9.00 in the evening – _____

i) 6.00 in the afternoon – _____

j) 8.00 in the morning – _____

18   Try Worksheet 4 *My day*.

19   Copy the time line, or use Worksheet 5.

Look at **Question 18** 'My day'.

Put the timetable for your day on a time line.

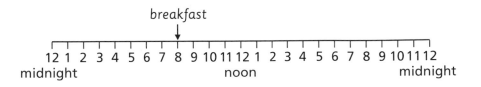

breakfast

12 1 2 3 4 5 6 7 8 9 10 11 12 1 2 3 4 5 6 7 8 9 10 11 12
midnight   noon   midnight

20 A school timetable got mixed up.

Activity	Time
lunch	12 noon
assembly	half-past nine
concert (for parents)	7.30
school starts	9.15
morning break	quarter past eleven
lesson 1	9.45
school finishes	four o'clock

Draw the table, putting the activities in order, **earliest time first.**

Write **a.m.** or **p.m.** after each time in the table.

21 a) Write the days of the week in order, Monday first.

> **Remember**
> There are 7 days in 1 week.

**Wednesday   Sunday   Thursday   Monday   Saturday
Tuesday   Friday**

b) The school week starts on _____.

c) There is no school on _____ and
_____.

If today is **Wednesday**:

d) yesterday was _____

e) tomorrow will be _____.

22   Look at the times on each pair of clocks.

Copy and complete, using **earlier** or **later**.

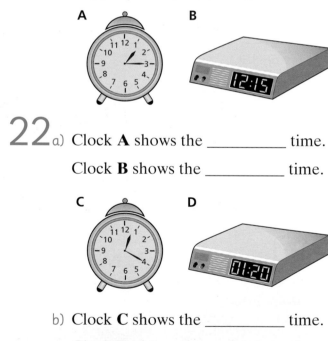

A   B

22 a) Clock **A** shows the _____ time.

Clock **B** shows the _____ time.

C   D

b) Clock **C** shows the _____ time.

Clock **D** shows the _____ time.

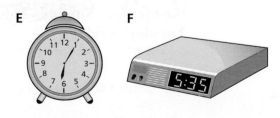

E   F

c) Clock **E** shows the _____ time.

Clock **F** shows the _____ time.

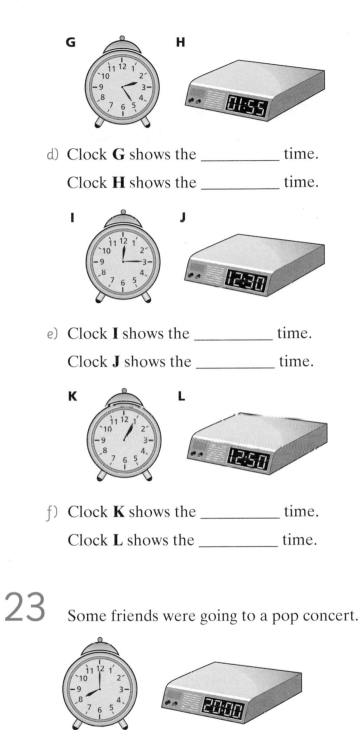

d) Clock **G** shows the _____ time.

Clock **H** shows the _____ time.

e) Clock **I** shows the _____ time.

Clock **J** shows the _____ time.

f) Clock **K** shows the _____ time.

Clock **L** shows the _____ time.

23 Some friends were going to a pop concert.

The train left at **8.00 p.m.**

They arrived at the station at these times.

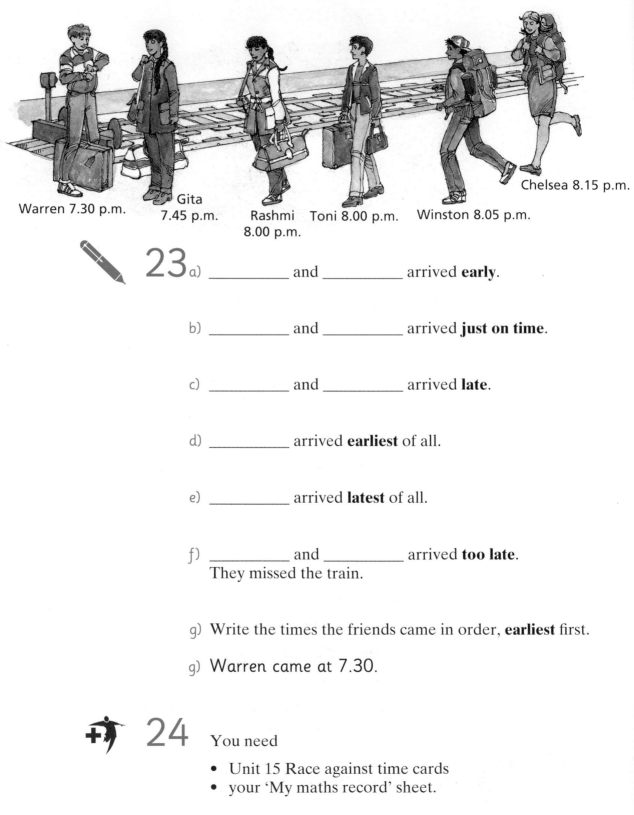

Warren 7.30 p.m.

Gita 7.45 p.m.

Rashmi 8.00 p.m.

Toni 8.00 p.m.

Winston 8.05 p.m.

Chelsea 8.15 p.m.

23

a) _____ and _____ arrived **early**.

b) _____ and _____ arrived **just on time**.

c) _____ and _____ arrived **late**.

d) _____ arrived **earliest** of all.

e) _____ arrived **latest** of all.

f) _____ and _____ arrived **too late**.
They missed the train.

g) Write the times the friends came in order, **earliest** first.

g) Warren came at 7.30.

24

You need

• Unit 15 Race against time cards
• your 'My maths record' sheet.

Race against time

1 Sort the race cards – this side up.

2 Take the cards 1 at a time.

3 Answer as quickly as you can.

4 Did you get it right?
   Look at the other side of the card.

15 minutes past 9
or quarter past 9

 5 When you get all the answers correct, ask a friend to test you.

6 Now **Race against time**

   Go for points!

   Ask your teacher to test and time you.

Remember
3 errors – 1 point
2 errors – 2 points
1 error – 3 points
0 errors – 5 points
Answer in 1 minute with 0 errors – 7 points

Now try Unit 15 Test.

Review 15

1 a) ☐ **halves** make **1 whole**.

b) ☐ **quarters** make **1 whole**.

c) ☐ **tenths** make **1 whole**.

2 Write as **decimals**.

a) $\frac{7}{10}$ = ☐.☐

b) $1\frac{3}{10}$ = ☐.☐

c) $\frac{2}{10}$ = ☐.☐

d) $1\frac{9}{10}$ = ☐.☐

3 Write as fractions.

a) 0.4 = ☐ tenths

b) 1.8 = ☐ whole one and ☐ tenths

c) 0.1 = ☐ tenth

d) 5.4 = ☐ whole one and ☐ tenths

4 Jade has 11p. Anne has 6p.

4 a) They have ☐p **altogether**.

b) _____ has the **most** money.
She has ☐p more.

c) _____ has the **least** money.
She has ☐p less.

5 a) **16** people make **2** teams of ☐ and ☐ left over.

b) **17** people make **2** teams of ☐ and ☐ left over.

6 There are ☐ half-litres in 1 litre.

7 Would you use a **metre**, a **kilogram** or a **litre** measure to find out how much milk to use in cooking?
I would use a _____ measure.